Contents

Preface

Pro/DESKTOP is a parametric feature-based solid modeling tool for making solid models of 3D objects, assemblies of solid models, engineering drawings from solid models and assemblies of solid models, and rendered images and animations of solids and assemblies of solid models. It uses three kinds of data file: you construct 3D solid parts and assemblies of solid parts in design files, engineering drawings of solids and assemblies of solids in drawing files, and rendered images and animations in photo album files.

Normally there are two stages in constructing a 3D solid part: analysis and synthesis. First you think about how you will decompose the 3D object into simple 3D solid features and combine the features to form the 3D object. Then you think about how to construct the features and recombine them accordingly. Basically, there are two kinds of solid features: sketched and placed. Sketched solid features are derived from sketches; you extrude, revolve, sweep, or loft the sketches to form sketched features. To establish planes for sketching, you construct workplanes. Placed solid features are preconstructed; you select a placed solid feature from the menu and specify the location and parameters.

An assembly is a collection of parts put together properly to serve a purpose. In the computer application, a design file depicting an assembly is linked to a set of design files corresponding to a set of solid parts.

There are three approaches to constructing an assembly: In the first approach, you design and construct all the solid parts in individual design files; then you start a design file depicting the assembly and place the solid parts in the assembly. In the second approach, you start an assembly design file; then you design and create the solid parts while you are working in the assembly environment, where you can perceive, visualize, and use the dimensions and features of the existing solid parts while designing new solid parts. The third approach is a hybrid approach; you design and construct some solid parts before constructing the assembly design file. Then you put the solid parts together in the assembly and design other solid parts in the assembly environment.

In the assembly, you can apply assembly constraints to align the parts and mate them together. When an assembly is complete, you can generate a parts list.

In a modern digital factory, you transmit digital design data about the solid parts and assemblies of components to the downstream operating departments. However, there are occasions when 2D engineering drawings are necessary for the purpose of communicating among the operators and designers. Constructing an engineering drawing consists of two tasks: generating engineering drawing views from a linked design file depicting an individual solid part or an assembly of solid parts, and adding annotations.

In addition to outputting engineering drawings, Pro/DESKTOP enables you to generate rendered images and animations from solid parts or assemblies of solid parts.

Organization of This Book

This book comprises seven chapters, with each chapter covering a major topic. Each chapter includes a summary and review questions. Tutorials are included in Chapters 2 through 7, so that you can practice the concepts learned in the chapter.

Chapter 1 provides a brief introduction to the functions of Pro/DESKTOP, explaining the three kinds of data files you will use. It also familiarizes you with the help system that assists you while you are designing.

Chapter 2 explains the key concepts of parametric feature-based solid modeling. You will learn how to construct parametric sketches and build four major kinds of sketched solid features. You will also learn how to combine features made from sketches and how to modify sketches and solid features.

Chapter 3 is the second part of solid modeling. You will learn how to drop library objects into your design, use preconstructed solid features in your design, and construct patterns and mirrors of features.

Chapter 4 introduces the concepts of assembly modeling and explains the three design approaches in making a design. You will learn how to put into an assembly a set of solid parts that you already constructed. You will also learn how to construct solid parts in the context of the assembly.

Chapter 5 deals with advanced modeling techniques: constructing a complex solid in a non-linear way, modifying a solid's face and body, treating imported surface objects, and using various kinds of design tools.

Chapter 6 delineates the concepts of engineering drafting. You will generate orthographic views from design files depicting a solid part or an assembly of solid parts and add annotations to a drawing.

Chapter 7 explains how to construct a rendered image and animation by using the photo album.

About the Companion CD-ROM

The companion CD-ROM found at the back of this book contains all Pro/DESKTOP files used in conjunction with exercises included in this book.

Acknowledgments

This book would never have been realized without the contribution of many individuals. Special thanks go to the professionals who reviewed the book.

Thomas Farmer, Oelwein High School, Oelwein, Iowa

Kenneth McDermith, Mohonasen High School, Schenectady, New York

Steve Ullrich, Hopkins High School, Minnetonka, Minnesota

Several people at Delmar Learning also deserve special mention, particularly James DeVoe, the senior acquisitions editor; John Fisher, the senior developmental editor who worked closely with me on this book; Alar Elken, Vice President Technology and Trades SBU; Katherine Bevington, editorial assistant; Stacy Masucci, the production editor; John Shanley of Phoenix Creative Graphics, the compositor; Agrawal Rajiv, the technical editor who reviewed the current manuscript in detail; and Gail Taylor, the copy editor. Special thanks go to Sky Sit of PTC who gave me much valuable advice.

Ron K. C. Cheng

Photo Album

To collate a set of images of solid parts and assemblies of solid parts for presentation and references, you construct a photo album. (See Figure 1–3.) You will learn how to construct photo albums in Chapter 7.

Figure 1–3
Photo album

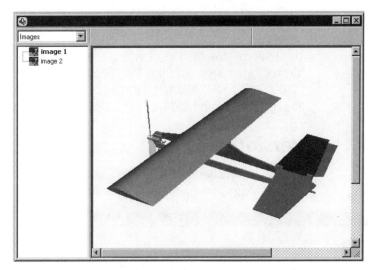

Data File and User Interface

Corresponding to the three main functions, Pro/DESKTOP has three kinds of data files, design file, drawing file, and photo album file, and one additional file type, design session. Their extensions are: *des*, *dra*, *alb*, and *ses*.

File Types

To start a new design file, engineering drawing file, or photo album file, you select Design, Engineering Drawing, or Photo Album from the New cascading menu of the File menu or select Design, Engineering Drawing, or Photo Album from the Standard toolbar. (See Figure 1–4.)

Figure 1–4
Starting a new design file, engineering drawing file, or photo album file

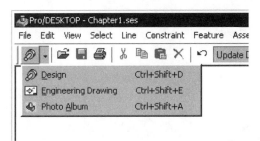

File Open and Save

To make use of design data produced by other computer-aided design tools, you can open file formats other than Pro/DESKTOP by specifying the file type in the Open dialog box. Once saved, the file will become a Pro/DESKTOP file. File formats that can be opened are Parasolid, IGES, STEP, SAT, VDA, AutoCAD DXF, and AutoCAD DWG.

To facilitate downstream computerized operations, you can save your file to formats other than Pro/DESKTOP formats, such as Parasolid, IGES, STEP, SAT, VDA, Medusa ASCII, Medusa Binary, ProductView, Stereo Lithography (STL), Jpeg, VRML, and Bitmap.

Versions

To help experiment with changes and safely return to a particular state of the design, you save your design in different versions by selecting File > Versions. In the Save Versions dialog box shown in Figure 1–5, select the Save Now button. In the Version dialog box that follows (shown in Figure 1–6), you can, optionally, write down a comment for later reference and select the OK button. To return to a certain version, select a version from the table and select the Restore button.

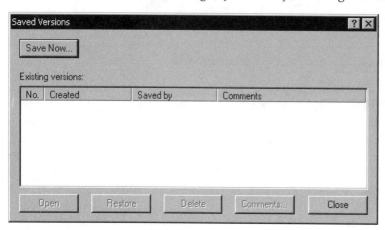

Figure 1–5 Save Versions dialog box

Figure 1–6
Version dialog box

Session

A session concerns the working environment, which is a collection of opened files, including window positions and window sizes. By opening a saved session, you can open all the related files in a single operation. This is particularly useful while you are working on a project. A session file has the *.ses* extension.

User Interface

Although the user interfaces of the data files are slightly different, they all have a working window, a menu, a palette, and a number of tool-bars, providing a working space and delineating different sets of commands for designated purposes.

Design File

We use the design file for two purposes: constructing 3D solid parts and assembling 3D solid parts. Figure 1–7 shows the user interface of a design file. Typically, a design file has a design window where you construct the solid part or the assembly, a palette at the left, and several toolbars located at the top, right, and bottom of the window.

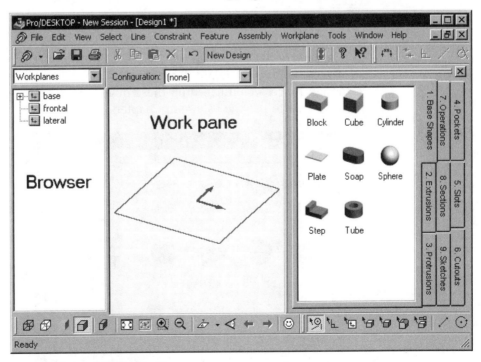

Figure 1–7 Design file

Design Window

The design window is used for constructing 3D components or an assembly of components. As shown in Figure 1–7, it has, in addition to the work pane, a dialog bar at the top and a browser pane at the left.

The work pane is the main working area of the design window. You can regard it as a viewing window through which you see the 3D space of unlimited size. As with a camera, you can zoom in, zoom out, and pan. At the top of the work pane is a dialog bar. From left to right, it has a browser list box, a configuration list box, and an area to provide feedback during sketching. The browser list box offers four options, enabling you to turn off the browser pane or to display the components browser, features browser, or the workplane browser. The configuration list box enables you to select one of the configurations of the design, if there is more than one design configuration. The prompt area provides clues or messages.

Depending on the selection in the browser list box of the dialog bar, the browser pane at the left of the design window shows objects in a design file in different formats. You can choose among four options (none, components, features, and workplanes) from the browser list box of the dialog bar.

Palette

The Palette bar shown in Figure 1–8 has a number of tabs in which various kinds of standard library objects are available. To speed up the design process, you can drop these objects into your design. You will learn how to use the Palette in Chapter 3.

Figure 1–8
Palette

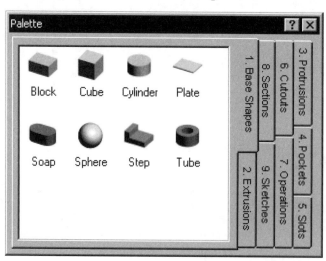

*Figure 1–9
Customize
dialog box*

Toolbars

Tools for constructing solid parts and assemblies of solid parts are available from the menu and the toolbars. To display the toolbars, select View > Customize and then check the appropriate boxes in the Customize dialog box. (See Figure 1–9.)

Engineering Drawing File

We use the engineering drawing file for constructing 2D engineering drawings of solid parts and assemblies of solid parts. Similar to the design file, the user interface of an engineering drawing file also has a palette and a number of toolbars, in addition to the engineering drawing window. Slightly different from the design window, the engineering drawing window also has a dialog bar, but it does not have different browsers or a configuration dropdown list. However, the dialog bar provides feedback during sketching. It only has a browser pane at the left and a drawing pane at the right. The browser pane shows the objects in an engineering drawing file and the drawing pane is a working space of unlimited size. (See Figure 1–10.)

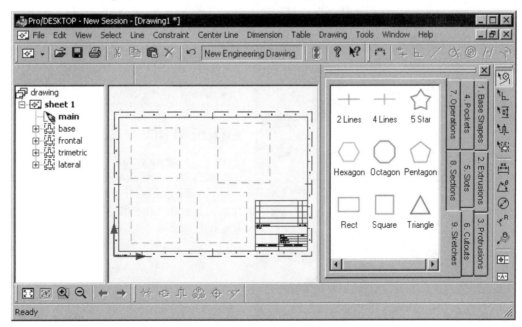

Figure 1–10 Engineering drawing file

Photo Album File

The photo album file is used for making photo albums of images of solid parts and assemblies of solid parts. As shown in Figure 1–10, it has an Album window and a number of toolbars, but the Palette is not applicable here. The Album window of a photo album file has three areas, the dialog bar at the top, the browser pane at the left, and the album pane at the right.

Figure 1–11
Photo Album file

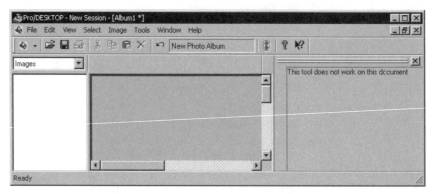

System Settings

Before starting to use Pro/DESKTOP as a tool to design and construct of 3D models, engineering drawings, and photo albums, take some time to study various system options to find how they affect your work. Select Tools > Options. In the Options dialog box, there are five tabs: Appearance, Performance, Directories, Units, and Drawing Standard.

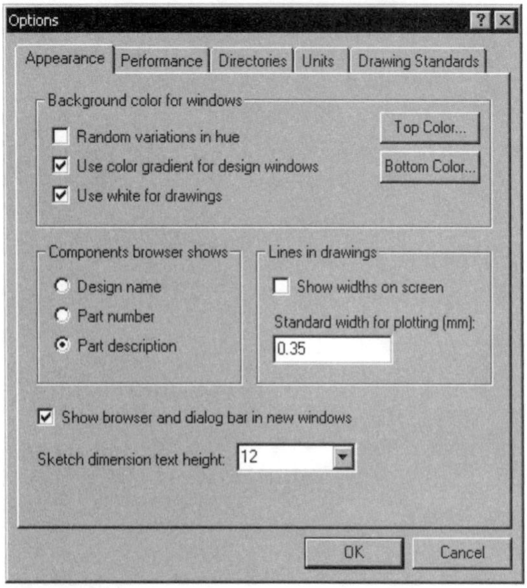

Appearance Tab

The Appearance tab sets the background colors of the windows, objects shown in the component browser, and lines in the drawing. It also decides the display of browser and dialog bar and determines the height of sketch dimensions. (See Figure 1–12.)

Figure 1–12 Appearance tab of the Options dialog box

Performance Tab

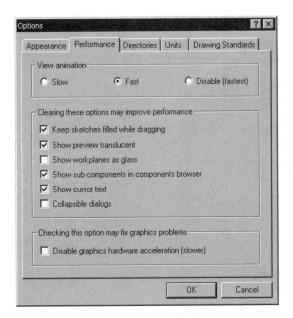

The Performance tab has three major areas: View animation, Performance improvement factors, and Graphics problem. In the View animation area, you can sets animation viewing speed. In the second area, there are check boxes that help improve performance. In the last area, there is a check box that helps fix graphics problems. (See Figure 1–13.)

Figure 1–13 Performance tab of the Options dialog box

Directories Tab

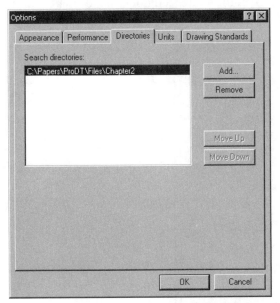

The Directories tab manipulates search directories. (See Figure 1–14.) The directories are searched in the specified order to resolve interfile dependencies. Interfile dependencies are created when you make an assembly of parts, an engineering drawing or a photo album.

Figure 1–14 Directories tab of the Options dialog box

Units Tab

The Units tab sets model and drawing units. (See Figure 1–15.)

Figure 1–15
Units tab of the
Options dialog box

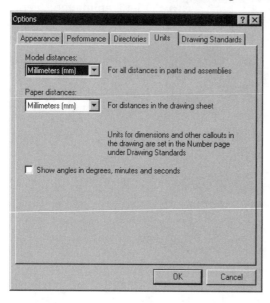

Drawing Standards Tab

The Drawing Standards tab manipulates engineering drawing standards deployed in a drawing. (See Figure 1–16.)

Figure 1–16
Drawing Standards
tab of the Options
dialog box

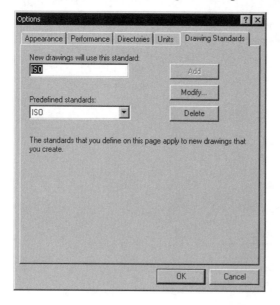

Help Systems

The comprehensive help system consists of several elements, available from the Help menu and the Standard toolbar.

Contents and Index

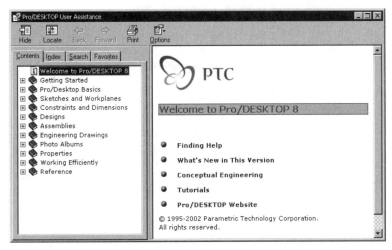

All information regarding the current installed Pro/DESKTOP application can be obtained by selecting Contents and Index from the Help menu. Figure 1–17 shows the Pro/DESKTOP User Assistance dialog box.

Figure 1–17 Pro/DESKTOP User Assistance dialog box

Context-Sensitive Help

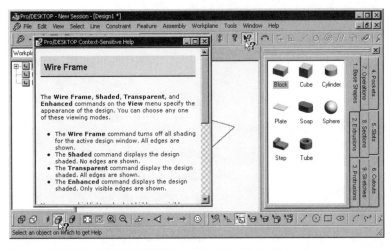

To find out specific information, select What's This? from the Help menu or the Context-Sensitive Help from the Standard toolbar and select an object or a menu command. Figure 1–18 shows the Pro/DESKTOP Context-Sensitive Help dialog box after the Design Window is selected. In addition, all dialog boxes offer context-sensitive help on every item inside the dialog box.

Figure 1–18 Pro/DESKTOP Context-Sensitive Help dialog box

Tutorials

To help familiarize yourself with Pro/DESKTOP, you can access a set of tutorials by selecting Tutorials from the Help menu. (See Figure 1–19.)

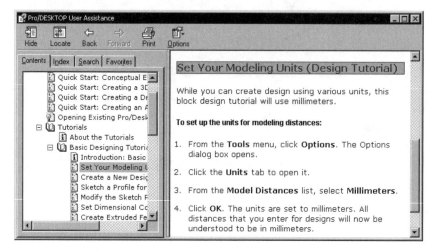

Tip of the Day

A randomized tip of the day is available by selecting Tip of the Day from the Help menu. (See Figure 1–20.)

PTC on the Web

Up-to-date information and help are available through the World Wide Web. You can connect to Web sites regarding product news, online support, and PTC by selecting Product News, Online Support, and PTC Home Page from the PTC on the Web cascading menu of the Help menu. (See Figures 1–21 through 1–23.)

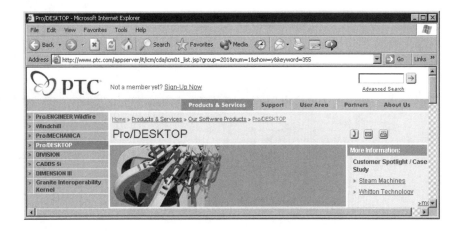

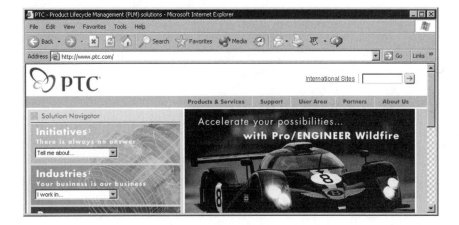

About Pro/DESKTOP

To learn more about the installed application, select About Pro/ DESKTOP from the Help menu. (See Figure 1–24.)

Figure 1–24
About Pro/DESKTOP

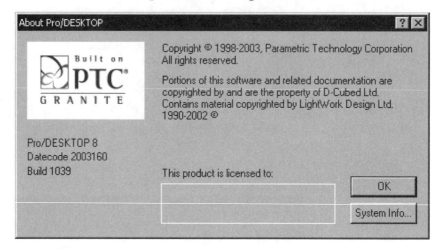

Summary

Pro/DESKTOP has three main functions, enabling us to construct solid parts and assemblies of solid parts, engineering drawings from solid parts and assemblies, and photo albums of images of solid parts and assemblies of solid parts.

Corresponding to its main functions, Pro/DESKTOP has three kinds of data files, namely, design file, engineering drawing files, and photo album files. Although the user interfaces of these data files are slightly different, they all have a working window, providing a working space.

To help you use the application, Pro/DESKTOP provides a comprehensive online help system.

Review Questions

1. What are the main functions of Pro/DESKTOP?

2. List the three kinds of Pro/DESKTOP data files and state their use.

3. What help elements are available?

CHAPTER 2

Solid Modeling I

Objectives

This chapter outlines the key concepts of the feature-based parametric approach in solid modeling, delineates four basic ways of constructing features of a solid model by sketching, details the ways to construct sketches and how to set up workplanes for sketching, and explains how to modify a solid model. After studying this chapter, you should be able to

❐ State the key concepts of the feature-based parametric modeling approach

❐ Construct extruded, revolved, swept, and lofted solid features by sketching

❐ Establish workplanes for sketching

❐ Modify a solid part

Overview

In the computer, a 3D solid model is an integrated mathematical representation depicting the vertices, edges, faces, and volume of the object it represents. Because any individual 3D object is unique in form and shape, it would not be feasible to derive a general mathematical expression to represent all kinds of 3D objects. Over the years, quite a number of mathematical methods of representing 3D objects in the computer have been developed. Among them, the feature-based parametric approach has become a commonly used 3D solid modeling method.

Using this method, you decompose a complex 3D solid part into simple features that can be represented in the computer, construct the features accordingly, and combine the features as you construct them. The basic way of constructing a solid feature is to construct a sketch or a number of sketches to depict the cross section(s) of the feature and select a command from the menu to produce a solid feature from the sketch(es).

In this chapter, you will learn how to construct sketches and use the sketches to construct four basic kinds of solid features. You will also learn how to establish workplanes for constructing sketches, ways to combine solid features, and method for redefining existing features. In Chapters 3 and 5, you will learn other techniques in solid modeling.

Feature-Based Parametric Modeling Concepts

Pro/DESKTOP is a feature-based parametric solid modeling application. To construct a 3D solid part, you think about how to decompose a complex object into simple unique features that can be constructed by using the tools provided by Pro/DESKTOP, construct the features one by one, and combine them as you construct them. Because Pro/DESKTOP is a parametric modeling system, you can redefine the parameters of the features any time during and after the construction of the solid model.

The Feature-Based Modeling Approach

The fundamental concept of making a 3D solid part by using the feature-based modeling approach is to use modular solid features as building block elements to construct 3D solids of complex shape. To make a 3D solid model this way, you go through two processes: The first is a top-down thinking process, and the second is a bottom-up construction process. To begin making a 3D solid part, you use deductive reasoning to analyze the solid part and decompose it into a number of simple and unique features definable by using the tools provided by Pro/DESKTOP, think about how to combine the features, and determine the sequence of construction. After this top-down thinking process, you perform the bottom-up construction operation to make the features one by one and combine them together as you construct them. (See Figure 2–1.)

Figure 2–1
Modular solid
features (left)
composing a solid
object (right)

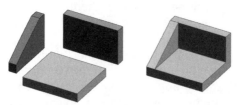

Modular solid features are constructed in two ways. In the first way, you construct a sketch or a number of sketches to depict the cross section of the features and select an appropriate command from the menu to make a solid feature from the sketches. In the second way, you select a pre-constructed solid feature from the menu and specify the parameters of the feature. Typical parameters are size and orientation. In this chapter, you will learn how to construct solid features by sketching. You will learn the second way and other modeling methods in Chapters 3 and 5.

Parametric Modeling System

When you construct a model in the computer, information regarding the form, shape, size, and related properties are called parameters. For example, we have to specify the diameter and height to construct a cylinder. The diameter and height here are the parameters of the cylinder. In a parametric system, all the parameters are modifiable any time during or after the creation of the features. Figure 2–2 shows the solid model modified with different parameters.

Figure 2–2 Changing the parameters

Because feature parameters are modifiable even after you have saved the file, closed it, and re-opened it, you do not have to worry too much about the exact dimensions of the features in the early stage of your design, when you only have an object's general form and shape. Naturally, you can deploy the computer as an electronic sketching board to record your initial design idea and concept. In the absence of definite dimensions, you can still proceed to construct a rough sketch or a number of sketches to depict the cross section(s) of a feature and use the appropriate command to construct a feature from the sketches. After making the feature by using some preliminary parameters, you can see better the outcome of construction. Thus, you can improvise your design by redefining the parameters. You will learn various ways to redefine the parameters of a solid part in this chapter.

Making Solid Features by Sketching

The fundamental way of constructing a 3D solid feature is to construct a sketch or a number of sketches to depict the cross section(s) of a solid feature, select a command from the menu, and let the computer produce a solid feature from the sketch(es). Using sketches, you can construct four basic kinds of solid features: extruded solid feature, revolved solid feature, swept solid feature, and lofted solid feature. Collectively, they are called sketched features. (See Figure 2–3.)

Figure 2–3
Four basic kinds of features constructed from sketches depicting their cross section

Sketching and Workplanes

To construct a sketch, you need a 2D plane, which can reside on one of the default workplanes, planar faces of an existing solid part, or user-defined workplanes. By default, there are three workplanes (base, frontal, and lateral) mutually perpendicular to each other, and there is a default sketch established on one of the default workplanes, the base workplane. After you start a new design file, you can start working on the default sketch or establish a sketch on one of the other two workplanes and construct a sketch there. Figure 2–4 shows a sketch constructed on the default workplane.

Figure 2–4
A sketch constructed on the base workplane

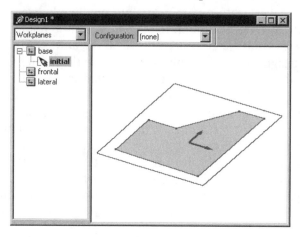

Extruded Solid Feature

An extruded solid feature is a solid feature with a uniform cross section. It is the simplest kind of sketched solid feature and is formed by first constructing a sketch profile on a 2D sketch plane and then extruding the profile in a direction perpendicular to the sketch plane. You can extrude the profile above the sketch plane, below the sketch plane, or symmetric about the sketch plane. In addition to extruding the entire closed profile as if it is a piece of lamina, you can treat it as a thin shell by selecting the thin option. Figure 2–5 shows four ways of extruding a profile.

Figure 2–5
From left to right:
extruding above,
below, and
symmetric about the
sketch plane, and as
thin object

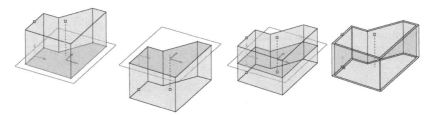

Projecting Profile

A derivative of the extrude profile command is the project profile command, which works similarly to the extrude profile command. Instead of extruding the sketch profile a specified distance, it projects the sketch profile to the next face, through the entire solid part, or to a selected face. Figure 2–6 shows a sketch profile projected to a selected face of the solid part. Like extruding a profile as a thin object, you can also project a profile to form a thin object.

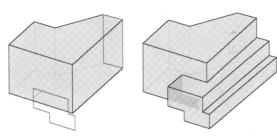

Figure 2–6 Projecting a sketch to a selected face of a solid part

Revolved Solid Feature

The second way of treating a sketch profile is to revolve it about an axis that is coplanar to the sketch. Using the revolve command, you can revolve the profile above, below, or symmetric about the sketch plane. (See Figure 2–7.) By selecting the thin option while revolving, you obtain a thin revolved object.

Figure 2–7
Revolving a sketch
above, below, and
symmetric about the
sketch plane

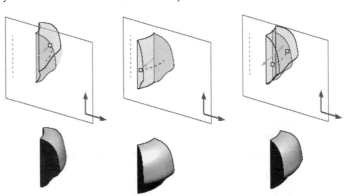

Swept Solid Feature

Sweeping is an operation in which a sketch profile is swept along another sketch, which we call the sketch path. The volume described by the sweeping action forms a swept solid feature. There are two kinds of swept solid features. The first kind consists of a sketch profile depicting the cross section and a sketch path depicting the sweeping path. The second kind also consists of a sketch profile describing the cross section, but the sketch path is not needed. Instead, a helical path is specified, and a helical swept solid feature in the form of a coil is constructed. (See Figure 2–8.)

Figure 2–8
Sweeping along a sketch path (left) and along a specified helical path (right)

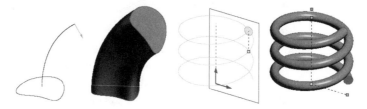

The thin object option is available for sweeping along a sketch path but not for the helical path.

Lofted Solid Feature

To construct a free-form 3D solid feature, you construct a number of sketch profiles on different sketch planes to depict the cross sections of the feature and loft through the sketches. (See Figure 2–9.) The thin object option is not available.

Figure 2–9
Lofting through a number of sketches

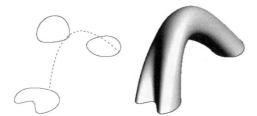

Thin Objects

Among the six kinds of sketched solid features mentioned above (extrude, project, revolve, sweep along sketch path, sweep along helical path, and loft), four of them provide the option of making a thin object in addition to making a solid object (extrude, project, revolve, and sweep along sketch path). By selecting the thin option and specifying the thickness while making the solid feature, you can construct a thin solid.

Combining Solid Features

In constructing a solid part that has more than one sketched solid feature, you have to decide how to combine the second and subsequent sketched solid features as you construct them. You can add the new feature to the solid part, subtract the new feature from the solid part, or intersect the new feature with the solid part.

Add Material

Using the Add Material option as you construct a sketched solid feature, you add the new feature to the existing solid part. The resulting solid part will have a volume enclosed by the existing solid part or the newly constructed sketched solid feature. Figure 2–10 shows a revolved solid feature (cone) added to an extruded solid feature (hexagonal prism).

Figure 2–10
Adding material
from the revolved
solid feature to
the extruded solid
feature

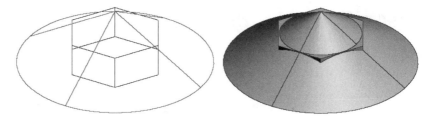

Subtract Material

Selecting the Subtract Material option while constructing a new sketched solid feature subtracts the new sketched solid feature from the existing solid part. The final solid part will have a volume consisting of the existing solid part less the volume enclosed by the new feature. Figure 2–11 shows a revolved solid feature subtracted from an extruded solid feature.

Figure 2–11
Subtracting material
from the revolved
solid feature to
the extruded solid
feature

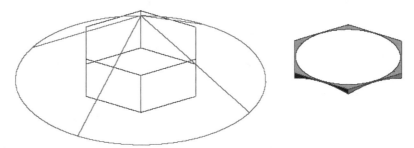

Intersect Material

The third option for combining is Intersect Material. It produces a solid part that has a volume common to the existing solid part and the new feature. See Figure 2–12 shows the revolved solid feature intersected with the extruded solid feature.

Figure 2–12
Intersecting material
from the revolved
solid feature with
the extruded solid
feature

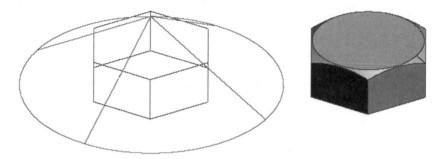

Selection Mode, Browser Pane, and Display Control

You need to be aware that, while making a solid model, from time to time you may have to select different kinds of objects for manipulation. For example, you may have to select a sketch element to drag it to a new location, and you may have to select a dimension to change its value property. Because there may be various kinds of objects displaying in the work pane, you have to pre-set the kind of objects to be selected by setting the object selection mode before actually selecting the object that you want.

A way to manipulate objects in a design file is to select them from the browser pane, which can be set to display components, features, or workplanes in a design file.

The work pane is a working window through which you see the objects constructed in the unlimited 3D virtual space. To help examine various objects here, you use display control tools.

Setting Selection Mode

You can set the selection mode by selecting an item from the Select menu or the Design toolbar. Figure 2–13 shows the selection setting mode commands available from the Design toolbar.

*Figure 2–13
Selection mode
setting from the
Design toolbar*

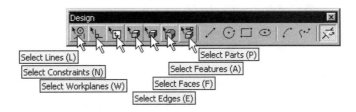

You can set selection mode to lines, constraints, workplanes, edges, faces, features, or parts. For example, you can set selection mode to lines while working on a sketch, and set selection mode to constraints while manipulating the dimension and geometric constraints of a sketch.

Using the Browser Pane

Because your concern here in this chapter is to construct solid parts, you can set the browser pane to display the objects in terms of the features in the solid part or the workplanes and the sketches on the workplanes. (See Figure 2–14.)

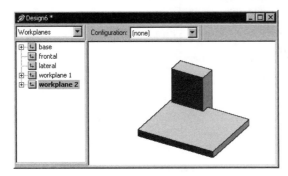 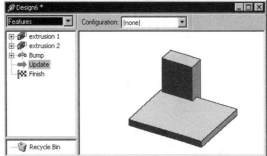

Figure 2–14 Browser pane set to display features (left) and workplanes (right)

Display Control

Display control tools are available from the View menu and the Views toolbar shown in Figure 2–15.

*Figure 2–15
Views toolbar*

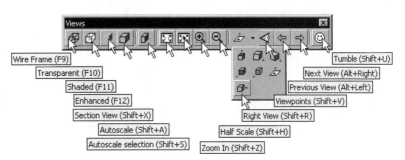

In addition to the menu and the toolbar, you can also use the keyboard, the mouse buttons, and the mouse wheel. Table 2–1 shows a comprehensive list of display control tools.

Table 2–1: Display control tools

Menu	View Toolbar	Keyboard and Mouse	Function
View > Autoscale	Autoscale	Shift + A	Fill the work pane with all the objects.
View > Autoscale Selection	Autoscale selection	Shift + S	Fill the work pane with selected object.
View > Half Scale	Half Scale	Shift + H	Reduce the current zoom scale by 50%.
View > Zoom In	Zoom In	Shift + Z	Fill the work pane with area specified by a rectangular bounding box.
—	—	Mouse Wheel or Control + Middle Mouse Button	Zoom in and zoom out dynamically.
—	—	Shift + Middle Mouse Button	Pan the display.
View > Wire Frame	Wire Frame	F9	Set the appearance of the design by not shading the faces and showing all the edges.
View > Transparent	Transparent	F10	Set the appearance of the design by shading the faces and turning on all the edges.
View > Shaded	Shaded	F11	Set the appearance of the design by shading the faces and turning off all the edges.
View > Enhanced	Enhanced	F12	Set the appearance of the design by shading the faces and turning on visible edges.
View > Go To > Isometric	View Isometric	Home or Shift + I	Set the display to an isometric view.
View > Go To > Trimetric	View Trimetric	End or Shift + T	Set the display to a trimetric view.
View > Go To > Plan	Plan View	Shift + P	Set the display to the plan view.
View > Go To > Front Elevation	Front View	Shift + N	Set the display to the front elevation.
View > Go To > Right Elevation	Right View	Shift + R	Set the display to the right elevation.

View > Go To > Onto Face	View Onto Face	Shift + F	Set the display to a selected face.
View > Go To > Onto Workplane	View Onto Workplane	Shift + W	Set the display to a selected workplane.
View > Go To > Previous	Previous View	Alt + Left	Set the display to the last view configuration.
View > Go To > Next	Next View	Alt + Right	Set the display to the next view configuration.
View > Rotate > Spin Left	—	Left	Spin the display to the left.
View > Rotate > Spin Right	—	Right	Spin the display to the right.
View > Rotate > Tilt Up	—	Up	Tilt the display up.
View > Rotate > Tilt Down	—	Down	Tilt the display down.
View > Rotate > Turn Counterclockwise	—	Page Up	Turn the display in a counterclockwise direction.
View > Rotate > Turn Clockwise	—	Page Down	Turn the display in a clockwise direction.
—	—	Middle Mouse Button	Rotate the display in all directions.
View > Viewports	Viewports	Shift + V	Manipulate viewpoints.
View > Section	Section	Shift + X	Set the display to a section view across a selected workplane.
View > Show Quilts	—	—	Displays surface quilts.
View > Tumble	Tumble	Shift + U	Spin the display in all directions.

Sketch and Workplanes

To construct a sketch for making a sketched solid feature, you use sketch tools. A sketch needs to be established on a workplane. Thus, you need to know how to construct workplanes.

Sketch Tools

The sketch tools available for making a sketch can be divided into two categories: tools for making sketch elements and tools for constraining the sketch elements.

Lines

In Pro/DESKTOP terms, sketch elements are called lines. You can access line construction tools from the Line menu or the Design toolbar. Figure 2–16 shows the line construction commands available from the Design toolbar.

Figure 2–16
Sketch element
construction tools on
the Design toolbar

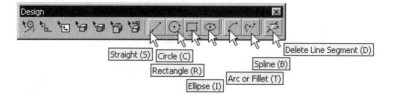

The tools for constructing sketch elements are listed in Table 2–2.

Table 2–2 Line tools

Menu	Design Toolbar	Function
Line > Straight	Straight	Construct a straight line.
Line > Circle	Circle	Construct a circle.
Line > Rectangle	Rectangle	Construct a rectangle.
Line > Ellipse	Ellipse	Construct an ellipse.
Line > Arc	Arc or Fillet	Construct an arc or a fillet between two straight lines.
Line > Spline	Spline	Construct a spline.
Line > Delete Segments	Delete Line Segment	Delete sketch element.
Line > Mirror	—	Construct a set of mirrored sketch elements about a mirror axis.
Line > Offset Chain	—	Construct offset sketch elements.
Line > Project	—	Project feature edges onto the current workplane.
Line > Convert To Straight	—	Convert selected arc or spline to a straight line.
Line > Add Text Outline	—	Construct text objects.
Line > Toggle Construction	—	Convert selected sketch element to construction lines.
Line > Toggle Sketch Filled	—	Fill or un-fill close-loop sketch.
Line > Toggle Sketch Rigid	—	Cause a sketch to be rigid or revert a rigid sketch to a normal sketch.

Constraints

Without any constraint being applied, sketch elements are free to translate. To restrict their free movement, you apply constraints to them.

There are two kinds of constraints: dimension constraint and geometric constraint. Dimension constraint governs the size, separation, or orientation of the sketch element. For example, diameter of a circle and length of a line are dimension constraints. By selecting and changing the dimension constraint's value property, you modify the size of the circle or the line. Geometric constraints concern how sketch elements are related to each other. For example, a collinear constraint causes two straight lines to be collinear, and a concentric constraint causes two arc or circles to be concentric.

Besides applying constraints, you can fix a sketch element so that it cannot be translated, and you can also set the sketch to be rigid so that the sketch elements are rigidly fixed among themselves. Constraint tools are accessible from the Constraint menu and the Constraint toolbar shown in Figure 2–17.

Figure 2–17
Constraint toolbar

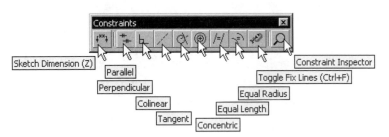

Tools for manipulating sketch element constraints are delineated in Table 2–3.

Table 2–3: Constraint tools

Menu	Design Toolbar	Function
Constraint > Dimension	Sketch Dimension	Construct constraint dimension.
Constraint > Parallel	Parallel	Set selected straight lines to be parallel.
Constraint > Perpendicular	Perpendicular	Set selected straight lines to be perpendicular to each other.
Constraint > Collinear	Collinear	Set selected straight lines to be collinear.
Constraint > Tangent	Tangent	Set selected line and arc or arc and arc to be tangent to each other.
Constraint > Concentric	Concentric	Set selected arc or circles to be concentric.
Constraint > Equal Length	Equal Length	Set selected straight lines to be equal in length.
Constraint > Equal Radius	Equal Radius	Set selected arcs to be equal in radius.
Constraint > Toggle Fixed	Toggle Fix Lines	Fix selected sketch elements.
Constraint > Toggle Reference	—	Toggle between a driving dimension and a reference dimension.
Constraint > Inspector	Constraint Inspector	Inspect and remove constraints already applied to selected sketch element.

Workplane Construction

In a design file, there are three default workplanes on which you can establish sketches. To construct additional workplanes, you use the New Workplane command accessible from the Workplane menu. Figure 2–18 shows the Workplane dialog box, providing six ways to construct a workplane: plane of object, offset, angled, tangent, plane through objects, and mid plane. (See Table 2–4.)

Figure 2–18 Workplane dialog box

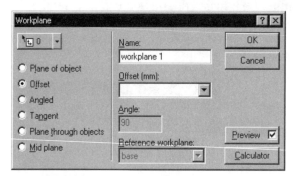

Table 2–4: Workplane Construction Methods

Method	Function
Plane of object	Construct a workplane coplanar with a selected face, edge, or workplane.
Offset	Construct a workplane offset from a selected face, edge, or workplane.
Angled	With reference to a workplane, construct a workplane at an angle to a selected edge, line, or an axis defined by a selected circular edge or a selected circular face.
Tangent	With reference to a workplane, construct a workplane tangent to a selected circular edge or circular face.
Plane through objects	Construct a workplane passing through selected objects, including coplanar lines, coplanar edges, and coplanar axes of cylinder, circle, or torus.
Mid plane	Construct a workplane midway between two selected workplanes or planar faces.

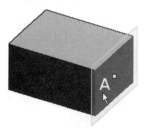

To construct a workplane on a plane of an object, select a planar face, a workplane, or a curved edge on a planar face. Figure 2–19 shows a workplane constructed on planar face A of a solid.

Figure 2–19 Workplane coplanar to a planar face of a solid

*Figure 2–20
Workplane offset
from selected
object's plane*

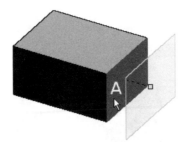

To construct an offset workplane, set an offset distance and select a planar face, a workplane, or a curved edge on a planar face. Figure 2–20 shows an offset workplane constructed from planar face A of a solid.

To construct an angular workplane, select a reference workplane from the Workplane dialog box and select a straight edge, an edge defining an axis, or a face defining an axis. Figure 2–21 shows an angular workplane passing through a selected edge and an angular workplane passing through the axis of a selected circular face.

*Figure 2–21
Workplane (with
reference to a
selected workplane)
at an angle to
selected object's
edge or axis*

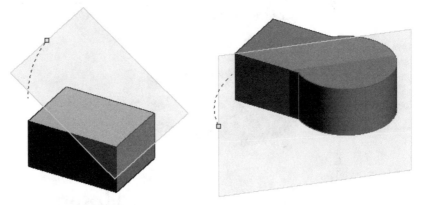

To construct a tangent workplane, select a reference workplane from the Workplane dialog box and select a circular edge or circular face. Figure 2–22 shows a tangent workplane tangent to a circular face and referenced to a workplane.

*Figure 2–22
Workplane tangent
to selected object's
edge or face*

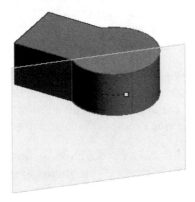

You can construct a workplane to pass through selected coplanar edges, circular faces of coplanar cylindrical objects, and faces of cylindrical and toroidal objects. (See Figure 2–23.)

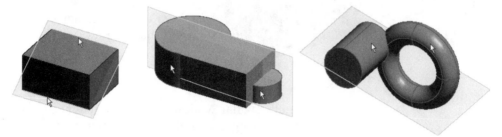

Figure 2–23 Workplane passing through selected objects

You can construct a workplane at mid plane between two selected planar faces or workplanes. (See Figure 2–24.)

*Figure 2–24
Workplane at mid
plane between two
parallel planar faces
or workplanes*

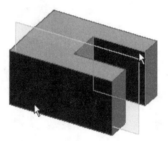

Solid Modeling Project—Ballpoint Pen

To reiterate, there are four major kinds of sketched solid features: extruded, revolved, swept, and lofted. In Pro/DESKTOP, there are two ways to construct extruded solid features and also two ways to construct swept solid features. Apart from extruding a profile, you can use the derivative of the extrude command, the project command, to project a profile to a selected termination face. To sweep a profile, you can sweep it along a path sketch or a specified helical path.

Among the four major kinds of solid features, extruded, revolved, and swept solid features have uniform cross sections defined by sketch profile. They are suitable for making objects with a uniform cross section. To construct a solid with varying cross sections, you use lofted solid features.

Before making a solid model, you have to think carefully about the form and shape of the features of the model and relate them to one of the sketched solid features.

Having decided which method(s) you will use to construct the features and the sequence of construction, you can start making the solid part. Naturally, construction of sketched features begins with making a sketch or a number of sketches depicting the cross section of the feature to be produced.

In the following, you will construct five component parts for making the ballpoint pen shown in Figure 2–25. While constructing these solid parts, you will learn various sketching methods, different selection modes, three methods of combining sketched solid features, and ways to construct workplanes.

Figure 2–25
Ballpoint pen project

Pen Tip

Figure 2–26 shows the pen tip of the pen. In reality, this part of the pen has several components: ball, ball tip, polymer tube, and ink. Here we will treat it as a single part and construct a dummy instead.

Figure 2–26
Pen tip

To construct this component, we will construct a sketch depicting the cross section along the axis of the component and revolve the sketch to a revolved solid. Now start a new design file.

1. Select File > New > Design or Design from the Standard toolbar.
2. Select Tools > Options.
3. In the Options dialog box, select the Units tab.
4. Set model distances to millimeters and select the OK button to close the Options dialog box.

Fixed Line

To prevent a line from accidentally moved from its designated position, you fix the line by selecting the line and then selecting Constraint > Toggle Fixed. Now construct a sketch on the default workplane. In the sketch, you will construct fixed lines and construction lines.

1. Select the + sign next to the Base workplane in the browser pane to expand it. You will find a default sketch (initial) activated. You will start working on this sketch.

2. Select View > Go To > Plan or select the Plan View button from the View toolbar.

3. Select Line > Straight or select the Straight button from the Design toolbar.

4. Move the cursor over the origin, shown at A in Figure 2–27.

5. When you see a black dot highlighted at the origin position of the coordinate icon, press and hold down the left mouse button.

6. Drag the mouse to the left and watch the reading at the cursor, showing the cursor's exact location.

7. Drag the mouse to point B (105,180°) and release the mouse button. (Note: To construct a horizontal line, you can hold down the SHIFT key.)

Figure 2–27
Horizontal line
being constructed

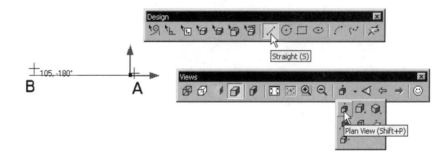

8. While the line is still selected, select Constraint > Toggle Fixed.

A horizontal straight line constructed from the origin to the left is fixed. Note that a small triangle is marked on the line, denoting that it is fixed. If you want to remove the fixed constraint, select the line and select Constraint > Toggle Fixed again.

Construction Line

While making a sketch, you can use construction lines to help establish relationships among sketch elements. Construction lines are shown as dashed lines. They are ignored in feature construction operations. There are three ways to convert a line to a construction line or revert a construction line to an ordinary line: select the line, right-click, and select Toggle Construction, select the line and press the CTRL key in conjunction with the G key, or select the line and select Line > Toggle Construction from the menu bar. Now construct a fixed construction line.

1. Select Line > Straight or select the Straight button from the Design toolbar.

2. Select end point A, hold down the mouse button, drag the mouse downward to point B (50,-90°), and release the mouse button. (See Figure 2–28.)

3. While the line is still selected, select Constraint > Toggle Fixed.

4. Right-click and select Toggle Construction.

Figure 2–28
Vertical line being
constructed

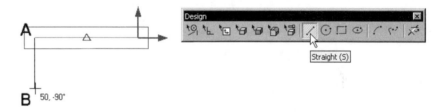

A fixed construction line is constructed. Note that a construction line is shown as a dashed line.

Sketching and Dimensioning

Now construct an arc tangent to the construction line.

1. Select Line > Arc or select the Arc or Fillet button from the Design toolbar.

2. Select line AB (Figure 2–29), if it is not selected.

3. Select end point B of line AB, drag the mouse to location C (exact location is unimportant), and release the mouse button. (Note: Care should be taken that the horizontal line is not highlighted. Otherwise, the command goes into the fillet mode. You should see that the cursor shows a free-standing arc, and not a fillet.)

4. If you used some other way to construct the arc and the arc is not tangent to line AB, hold down the SHIFT key, select line AB and arc BC, release the SHIFT key, and select Contraint > Tangent or select Tangent from the Constraint toolbar.

5. If tangent constraint already exists, a dialog box will appear. Select the OK button. (See Figure 2–30.)

Figure 2–29
Tangent arc being
constructed

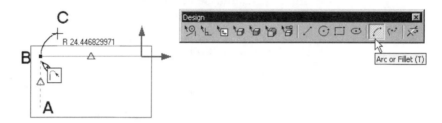

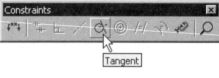

Figure 2–30 Tangent constraint and warning dialog box

Now construct three line segments.

1. Select Line > Straight or select the Straight button from the Design toolbar.

2. Select end point B (Figure 2–31) of the arc, drag the mouse to location C (exact location is unimportant), and release the mouse button.

3. Select end point C, drag the mouse to location D, and release the mouse button.

4. Select end point D, drag the mouse to location E, and release the mouse button.

5. Select arc AB and line BC and select Constraint > Tangent or select Tangent from the Constraint toolbar.

6. Pick a point anywhere in the empty space of the work pane to de-select the objects you already selected.

7. Select lines CD and AF and select Constraint > Parallel or select Parallel from the Constraint toolbar.

8. De-select the objects already selected, select lines CD and DE, and select Constraint > Perpendicular or select Perpendicular from the Constraint toolbar.

Figure 2–31 Lines being constructed

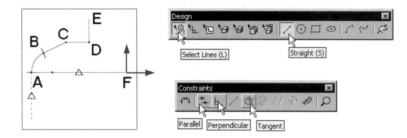

Now add dimensions to the sketch.

1. Select Constraint > Dimension or select Sketch Dimension from the Constraint toolbar.

2. Select line AB. Then move the mouse over line CD, drag the mouse to location E (Figure 2–32), and release the mouse button. (Note: You can pre-select line AB before selecting the command.)

3. Select the dimension and double-click. (Note: You have to set the selection mode to constraint prior to selecting and double-clicking the dimension.)

4. In the Properties dialog box, change the angle value to 20 and select the OK button.

5. Select Constraint > Dimension or select Sketch Dimension from the Constraint toolbar.

6. Select arc AC (Figure 2–32), drag the mouse starting from the arc AC to location F (Figure 2–32), and release the mouse button. This will place a diameter size dimensional constraint on the arc.

7. Double-click the dimension and change the dimension's value to 1.2 in the Properties dialog box. (Note: Selection mode needs to be set to constraint prior to double-clicking the dimension.)

8. Select Constraint > Dimension or select Sketch Dimension from the Constraint toolbar.

9. Select line GH (Figure 2–32) and point A (Figure 2–32), drag the mouse to location J (Figure 2–32), and release the mouse button. (Note: You may need to zoom into the sketch.)

10. Double-click the dimension and change the dimension's value to 8 in the Properties dialog box.

11. Select Constraint > Dimension or select Sketch Dimension from the Constraint toolbar.

12. Select lines DG and AB (Figure 2–32), drag the mouse to location F (Figure 2–32), and release the mouse button.

13. Double-click the dimension and change the dimension's value to 1.25 in the Properties dialog box.

14. Select Constraint > Dimension or select Sketch Dimension from the Constraint toolbar.

15. Select line AB (Figure 2–32), drag the mouse to location K (Figure 2–32), and release the mouse button.

16. Double-click the dimension and change the dimension's value to 105 in the Properties dialog box.

Several dimensions are constructed. (See Figure 2–33.)

Figure 2–32
Angular dimension
being constructed

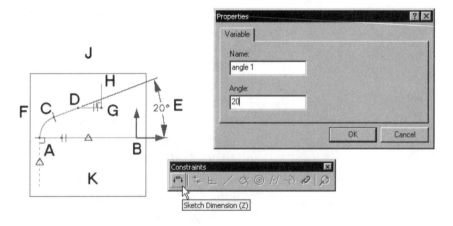

Figure 2–33
Dimensions added

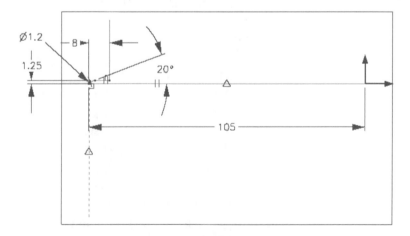

Now construct a few more lines to form a closed profile.

1. With reference to Figure 2–34, construct lines AB, BC, CD, DE, EF, and FG. By default, a closed profile is filled. If you do not want to fill the closed profile, select Line > Toggle Sketch Filled. If you want to fill the profile, select Line > Toggle Sketch Filled again. (Note: Line AB starts from point H of the previous figure. To construct these horizontal and vertical lines, you should hold down the SHIFT key.)

2. De-select objects already selected, select lines AB, CD, EF, and GH, and select Constraint > Parallel or select Parallel from the Constraint toolbar.

3. De-select the objects already selected, select lines BC, DE, and FG, and select Constraint > Parallel or select Parallel from the Constraint toolbar.

4. De-select the objects already selected, select lines FG and GH, and select Constraint > Perpendicular or select Perpendicular from the Constraint toolbar.

5. Use the dimension command to set the length of CD to 1.5, distance between line BC and point H to 32, distance between lines EF and HG to 1.75, and distance between lines CD and HG to 2.25. Also set lines AB and EF to be collinear. (See Figure 2–35.)

The profile is complete.

*Figure 2–34
Closed profile
constructed*

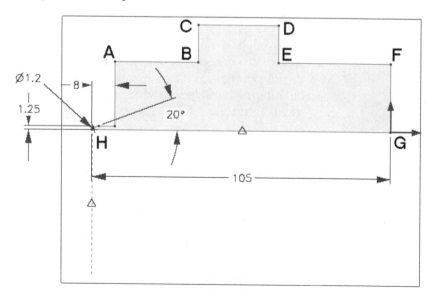

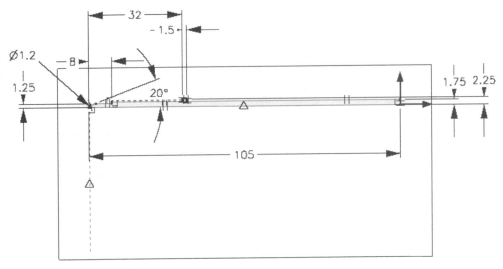

Figure 2–35 Dimensions added

Toggle Sketch Filled

By default, a closed sketch profile is filled with shaded color in your screen, denoting that it is a closed loop and is ready for further feature creation process. If you do not want a closed profile to be filled, perhaps for the purpose of clarity, select Line > Toggle Sketch Filled. If you decide to fill it again, select Line > Toggle Sketch Filled once more.

Toggle Sketch Rigid

If a sketch is not fully constrained, selecting a line element of the sketch and dragging it may change the shape of the sketch. By setting the line as rigid, you can drag the entire sketch to a new position without altering its shape. To set a sketch rigid, select Line > Toggle Sketch Rigid. To reset a rigid sketch, select Line > Toggle Sketch Rigid again.

Constructing a Revolved Solid Feature

Now revolve the sketch.

1. Select View > Go To > Trimetric or select View Trimetric from the View toolbar.
2. Select Feature > Revolve Profile or select Revolve Profile from the Features toolbar.
3. Select edge A indicated in Figure 2–36 to use it as the axis of revolution.

4. In the Revolve Profile dialog box, select the Add Material option, if it is not already selected. Note that you cannot choose Subtract Material or Intersect Material because this is the first feature of the solid part.

5. Set the angle of revolution to 360° and select the OK button.

The revolve solid is complete. (See Figure 2–26.) Save and close your file (file name: *BallPen01.des*). (Note: You should set up a folder in your computer and save your file there. Alternatively, you can simply save it in the My Documents folder.)

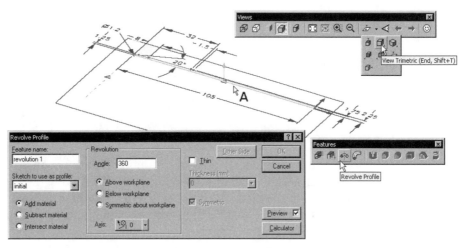

Figure 2–36 Revolve solid being constructed

Button

Figure 2–37 shows the ballpoint pen's button. To make this model, you will construct two extruded solid features.

Figure 2–37 Button

Now start a new design file and set the unit of measurement.

1. Select File > New > Design or Design from the Standard toolbar.

2. Set the unit of measurement to mm.

3. Set the display to the plan view.

Dropping Fixed Construction Lines from the Palette and Sketching

The palette shown in Figure 2–38 is a standard library consisting of various library objects, details of which will be covered in the next chapter. Here you will use the objects in the sketch tab. If the palette is not displayed on your screen, select Tools > Palette.

Figure 2–38
Palette

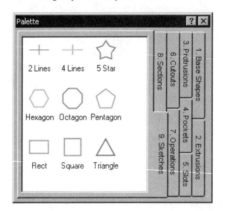

The "9. Sketches" tab has a number of sketches that you can drop into your design by double-clicking on them. Among the sketch objects, the "2 Lines" sketch contains a fixed horizontal construction line and a fixed vertical construction line. Dropping it into your design saves you time in constructing fixed construction lines.

Now drop two fixed construction lines from the palette.

1. Double-click "2 Lines" from the "9. Sketches" tab of the palette. Two fixed construction lines are dropped onto the **base** workplane and a new sketch named "2 Lines" gets created. These lines, passing through the origin, are construction lines and are fixed.

Now construct a number of line segments.

2. With reference to Figure 2–39, construct the remaining lines of the sketch. (Note: Remember to hold down the SHIFT key while constructing horizontal and vertical lines.)

3. Set lines AB, CD, EF, GH, JK, and LM to be parallel to the fixed construction line Q. (Note: You might get a message saying that "constraint already exists between the selected objects." If so, simply select the OK button.)

4. Set lines BC, DE, FG, HJ, KL, and MA to be parallel to the fixed construction line P.

5. Add dimensions accordingly.

6. Set lines MA to be collinear with the fixed construction line P.

7. Add a dimension to set the distance between line AB of Figure 2–39 and the fixed horizontal line to 2.4 mm. (See dimension X of Figure 2–40.)

The sketch is complete.

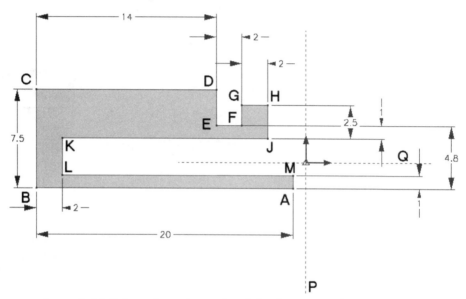

Figure 2–39 Palette object dropped and sketch constructed

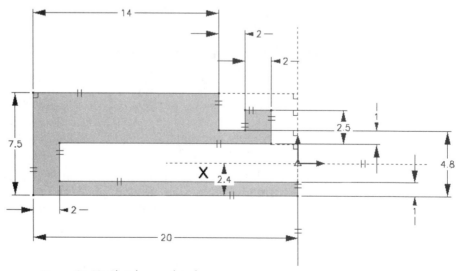

Figure 2–40 Sketch completed

Extruding the Profile

Now extrude the sketch in two directions symmetric about the workplane.

1. Select View > Go To > Trimetric or select View Trimetric from the View toolbar.

2. Select Feature > Extrude Profile or select Extrude Profile from the Feature toolbar.

3. In the Extrude Profile dialog box, set the extrusion distance to 2 mm, check the Symmetric about workplane option, and select the OK button. (See Figure 2–41.)

Figure 2–41
Profile being
extruded

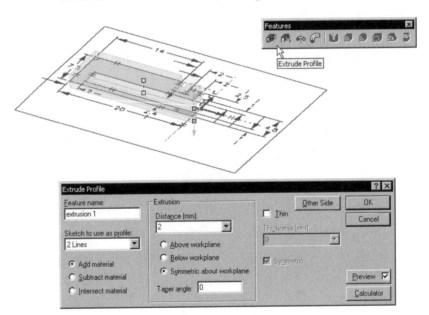

Second Extruded Solid Feature

Now construct the second extruded solid feature.

1. Select the lateral workplane from the browser pane, right-click, and select New Sketch. (See Figure 2–42.) (If the browser pane is not displaying the workplanes, select Workplane from the browser list.)

2. With reference to Figure 2–43, construct a circle with the center point located at the origin of the work axes. (Note: The origin is the current origin of the work axes.)

3. Add a dimension to set the size of the circle to 5 mm diameter.

4. Extrude the sketch a distance of 16 mm. (See Figure 2–44.)

The solid model is complete. (See Figure 2–37.) Save and close your file (file name: *BallPen02.des*).

Figure 2–42
New sketch being
constructed

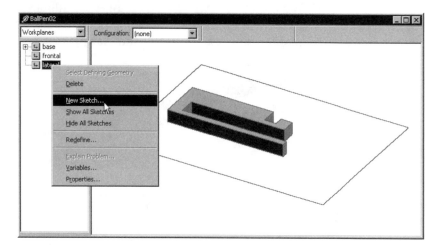

Figure 2–43
Circle constructed

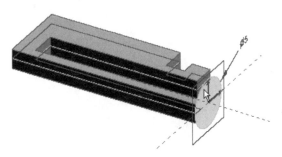

Figure 2–44
Profile being
extruded

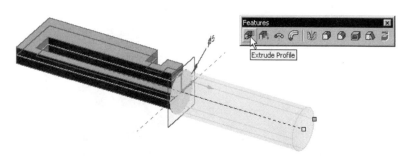

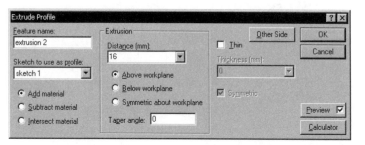

Barrel

Figure 2–45 shows the barrel of the ballpoint pen. To produce this component, you will construct a revolved solid feature, two projected solid features, and a swept solid feature.

Figure 2–45
Barrel

Revolved Solid Feature

Now start a new design file and construct a revolved solid feature.

1. Select File > New > Design or Design from the Standard toolbar.

2. Set the unit of measurement to mm.

3. Set the display to the plan view.

4. In accordance with Figure 2–46, drop two fixed construction lines onto the default workplane and make a closed profile. In the sketch, lines AJ, CD, EF, and HG are parallel to the horizontal fixed construction line X and lines AB, DE, FG, and HJ are parallel to the vertical fixed construction line Y.

5. Set line FG to be collinear with the vertical construction line Y.

Figure 2–46
Fixed construction lines dropped and closed profile constructed

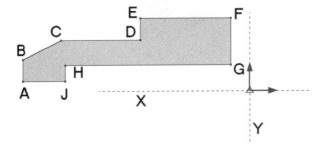

6. With reference to Figure 2–47, add dimensions to the sketch.

7. Set the display to trimetric.

8. Revolve the sketch 360° about line A indicated in Figure 2–48.

The revolved solid feature is constructed.

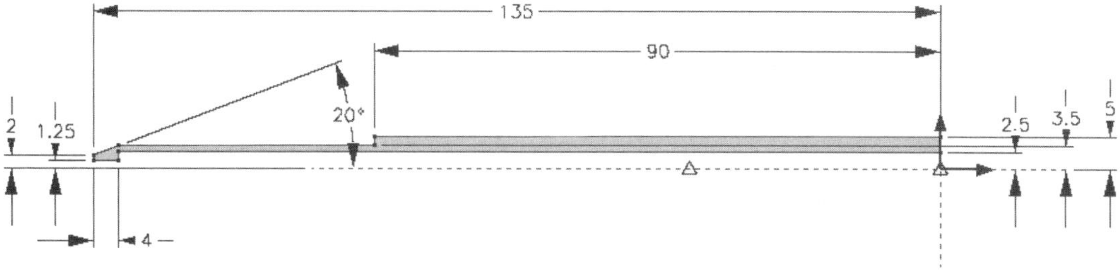

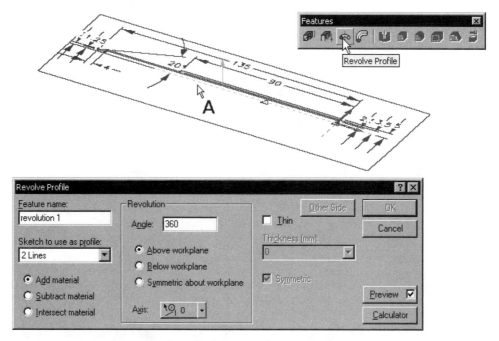

Figure 2–47 Dimensions added to the sketch

Figure 2–48 Sketch profile being revolved

Projected Solid Features

The project profile command is a derivative of the extrude profile command. In essence, it also extrudes a profile. However, instead of extruding the profile a specified distance, it extrudes the profile to the next face, through the entire solid part, or to a selected workplane. Apart from this, it also differs from the extrude profile command in that there are only two options (Add Material and Subtract Material) instead of three options. The Intersect Material option is not available. Now construct two projected solid features.

1. Select the lateral workplane from the browser pane, right-click, and select New Sketch. A new sketch is constructed on the workplane.

2. With reference to Figure 2–49, construct two construction lines and a rectangle.

Figure 2–49
New sketch
constructed on the
lateral workplane

Figure 2–50
Sketch constructed

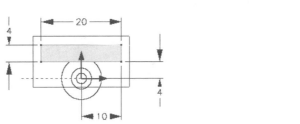

3. Set the display to right elevation and add dimensions to the sketch in accordance with Figure 2–50.

4. Set the display to trimetric and select Feature > Project Profile or select Project Profile from the Feature toolbar.

5. In the Project Profile dialog box, select Subtract material, set Projection Below workplane, set Extent to Thru entire part, and select the OK button.

Figure 2–51
Profile being
projected to subtract
the solid part

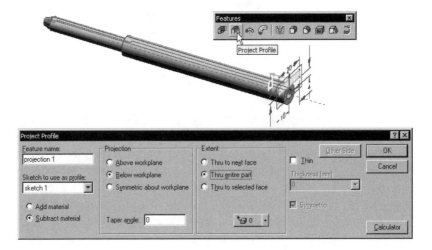

6. Set the selection mode to faces, select face A, right-click, and select New Sketch. (See Figure 2–52.)

7. Select line A of Figure 2–53, right-click, and select Project. The selected line is projected onto the active sketch.

8. Select the projected line B of Figure 2–53, right-click, and select Toggle Construction to change the projected line to a construction line.

9. With reference to Figure 2–53, construct two rectangles.

Figure 2–52
New sketch being
constructed

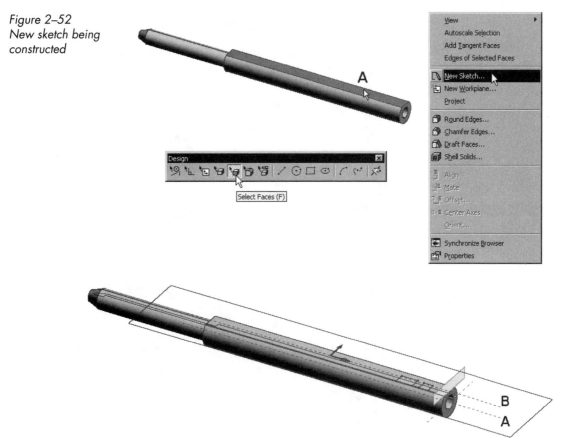

Figure 2–53 Sketch being constructed

10. Set the display to plan view and add dimensions to the sketch in accordance with Figure 2–54. (Note: The horizontal lines of the rectangles are collinear to each other.)

11. Set the display to trimetric.

12. Select Feature > Project Profile or select Project Profile from the Feature toolbar.

13. In the Project Profile dialog box, select Subtract material, set Projection to Below workplane, set Extent to Thru to next face, and select the OK button.

The profile is projected.

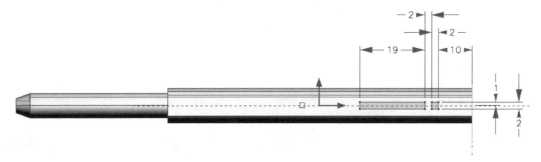

Figure 2–54 Sketch dimensioned

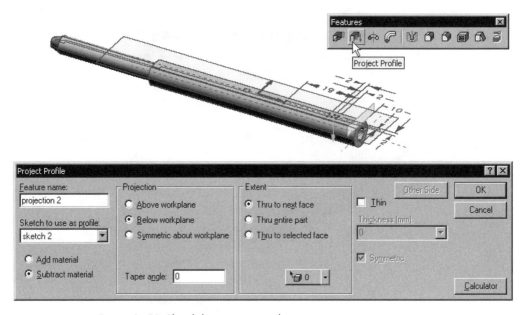

Figure 2–55 Sketch being projected

Swept Solid Feature

Basically, a sweeping operation requires two sketches residing on two workplanes that are mutually perpendicular to each other, using one of the sketches as the sweeping profile and the other sketch as the sweeping path. While sweeping, the sketch profile sweeps along the sketch path, describing a volume. A swept solid is best suited for making pipes, routes, or rails.

While a sketch profile has to be a close-loop profile, similar to what you use in extruding and revolving, a sketch path does not need to be a close-loop sketch. However, it is essential that the starting end of the sketch path terminate at the sketch plane on which you construct the sketch profile, and its tangent direction has to be perpendicular to the sketch profile.

Now hide all the sketches and construct a workplane for making a profile sketch of the swept solid feature.

1. From the Workplane menu, select Hide Other Sketches to hide all the sketches except the active one.

2. Select Workplane > New Workplane.

3. In the Workplane dialog box, select the Offset option, set selection mode to Faces from the dropdown selection control in the Workplane dialog box, and set the offset distance to –45.

4. Select face A indicated in Figure 2–56.

5. Select the OK button. A workplane is constructed.

Figure 2–56 Offset workplane being constructed

Now construct a profile sketch on the new workplane.

1. Select workplane B indicated in Figure 2–56, right-click, and select New Sketch.

2. Set the display to right elevation.

3. Select workplane A indicated in Figure 2–57, right-click, and select Project.

4. Select Line > Toggle Construction to change the projected line to a construction line.

5. In accordance with Figure 2–58, construct an ellipse and add dimensions. The center of the ellipse should coincide with the projected construction line. (Note: To select the center point of the ellipse, place the cursor inside the ellipse and near to the circumference. To select the endpoints of the ellipse, select the ellipse's quadrant position.)

Figure 2–57
Projecting the
workplane as a line
onto the sketch

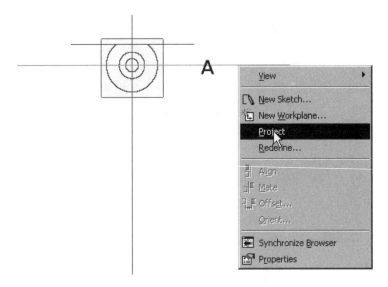

Figure 2–58
Sketch constructed

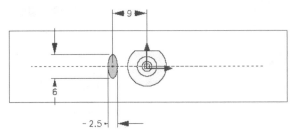

Now construct a sketch for the path of the swept solid feature.

1. Select Workplane > New Workplane.

2. In the Workplane dialog box, select Plane of object.

3. Select Base workplane from the browser pane.

4. Select OK button from the Workplane dialog box. A workplane is constructed.

5. Select the new workplane (which is defined by the base workplane) from the browser pane, right-click, and select New Sketch.

6. Set the display to the plan view.

7. With reference to Figure 2–59, select workplane A, right-click, and select Project.

8. Select Line > Toggle Construction to change the projected line to a construction line.

Figure 2–59
Workplane on
base workplane
constructed and line
being projected

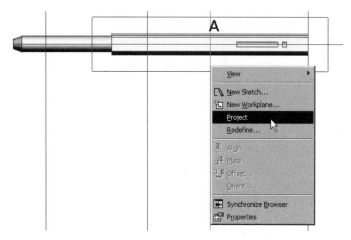

9. With reference to Figure 2–60, construct a line and an arc. Line B's endpoint should coincide with the construction line A, lines A and B should be perpendicular to each other, and line B should be tangent to arc C.

10. Add dimensions to the sketch in accordance with Figure 2–61. (Note: The 6.5 dimension is done by selecting line A and quadrant B and the 3.5 dimension is done by select endpoint C and circular edge B.)

Now construct a swept solid feature.

1. Select Feature > Sweep Profile > Along Sketch Path or select Sweep Profile Along Profile of the Feature toolbar.

2. In the Sweep Profile dialog box, select the sketches for the profile and the path from the pull-down list box. Use the ellipse as profile and the line and arc as the path. (Note: You can rename the sketches with meaningful names so that you can refer to them in the future.)

3. Select the OK button. (See Figure 2–62.)

Figure 2–60
Line and arc
constructed

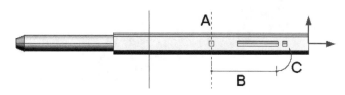

Figure 2–61
Dimensions added
to the sketch

Figure 2–61
Dimensions added
to the sketch

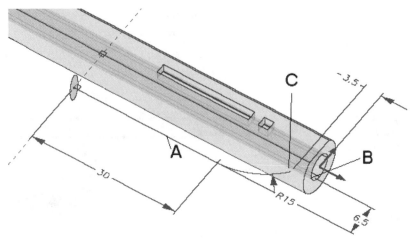

Figure 2–62
Swept solid being
constructed

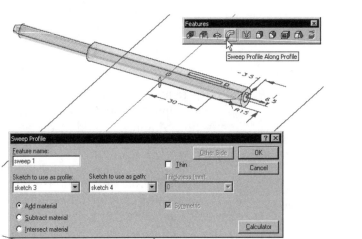

The barrel of the ballpoint pen is complete. (See Figure 2–45.) Save and close your file (file name: *BallPen03.des*).

Spring

Figure 2–63 shows the spring of the ballpoint pen. Basically, it is a helical swept solid. To facilitate assembly of the components, you will include two revolved solid features with the solid part.

Figure 2–63
Spring

Sweep Profile Along Helical Path

The second way of making a swept solid feature is to sweep a sketch pro-file along a helical path. Because a helical path can be expressed by using standard mathematical equations, you do not have to construct a helical sketch in order to construct a helical swept solid feature. Instead, you need only to select an axis for and specify the parameters of the helix. The helix axis can be a line or an edge coplanar with the sketch profile.

In addition to not requiring to construct a path sketch, a helical swept feature differs from an ordinary swept solid in that there are only two combination options: Add Material and Subtract Material. The Inter-sect Material option is not available.

Now start a new design file and construct a sketch for making the heli-cal swept solid.

1. Select File > New > Design or Design from the Standard toolbar.

2. Set the unit of measurement to mm.

3. Set the display to the plan view.

4. With reference to Figure 2–64, construct a vertical fixed construction line, a horizontal fixed construction line, and a circle with its center point coincident with the vertical construction line.

Figure 2–64
Fixed construction
lines and circle
constructed

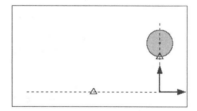

5. Add dimensions to the sketch in accordance with Figure 2–65. Note that the length of the horizontal line is used to define the length of the axis of the helix.

Figure 2–65
Dimensions added

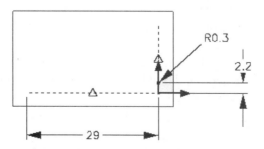

Now construct a swept solid feature.

1. Set the display to trimetric.

2. Select Feature > Sweep Profile > Along Helix.

3. Select the horizontal line as the axis of the helix.

4. In the Helical Sweep dialog box, set the pitch to 1 mm and select the OK button.

A helical swept solid is constructed. (See Figure 2–66.)

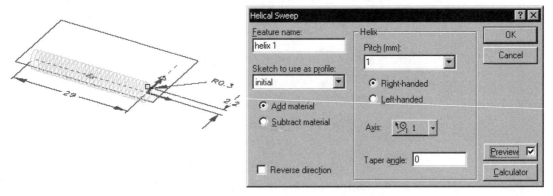

Figure 2–66 Helical swept solid being constructed

Reusing a Sketch

A sketch can be reused for different feature construction processes. Now you will reuse the same sketch that was used for making the swept solid feature to construct a revolved solid feature and add the revolved solid feature to the solid part.

1. Zoom the display with reference to Figure 2–67.

2. Select Feature > Revolve Profile or select the Revolve Profile button from the Feature toolbar.

3. Select the horizontal construction line (A of Figure 2–67) as the axis of rotation.

4. In the Revolve Profile dialog box, select Add material, set angle of revolution to 90, and select the OK button.

A revolved solid feature is constructed. This revolved solid feature's axis will be used in assembly of the ball pen components.

*Figure 2–67
Revolved solid
feature being
constructed*

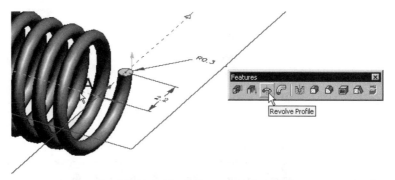

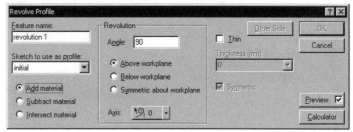

Now you will construct a revolved solid feature to intersect with the solid part.

1. Select the base workplane from the browser pane, right-click, and select New Sketch.

2. In the New Sketch dialog box, select the OK button.

3. With reference to Figure 2–68, construct a rectangle and add dimensions to it. Note that endpoints A and B of the rectangle should coincide with the endpoints of the horizontal construction line and dimension C's value is unimportant, as long as it is greater than the radius of the helix plus the radius of the profile of the helix.

*Figure 2–68
Sketch constructed*

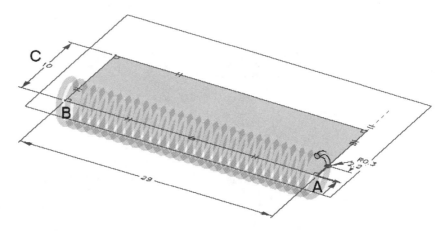

4. Select Feature > Revolve Profile or select the Revolve Profile button from the Feature toolbar.

5. Select the bottom line A of the rectangle as the axis of rotation. (See Figure 2–69.)

6. In the Revolve Profile dialog box, select Intersect material, set angle of revolution to 360, and select the OK button.

A revolved solid is constructed and intersected with the solid part. As a result of intersection, two flat faces are formed at the ends of the helical spring. These two faces will be used in assembling the spring. (Please refer to Figure 2–63.)

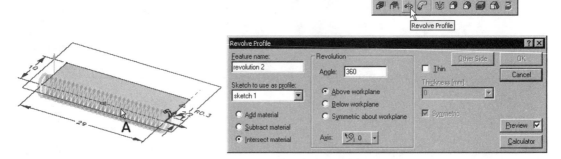

Figure 2–69 Revolved solid being constructed

The solid part is complete. Save and close your file (file name: *BallPen04.des*).

Hand Grip

Figure 2–70 shows the handgrip of the ballpoint pen. Its outer face is constructed by making a lofted solid feature and its core is an extruded solid.

*Figure 2–70
Hand grip of the
ball pen*

Loft Through Profiles

A lofted solid feature requires two or more sketch profiles to depict the cross sections of the feature. Depending on the complexity of the solid feature to be produced, you can construct as many sketch profiles as you like. Among the four kinds of sketched solid features, the lofted solid is the most complex one, in terms of form and shape. Therefore, it is best suited for making solids of free-form shape.

Now start a new design file and construct three sketches.

1. Select File > New > Design or Design from the Standard toolbar.

2. Set the unit of measurement to mm.

3. Select the lateral workplane from the browser pane, right-click, and select New Sketch.

4. In the New Sketch dialog box, select the OK button.

5. Select Workplane > Delete Empty Sketches. The default sketch on the base workplane is deleted.

6. Double-click 2 Lines from the "9. Sketches" tab of the Palette to drop a sketch with a fixed horizontal construction line and a fixed vertical construction line.

7. Set Line > Spline or select the Spline button from the Design toolbar.

8. Select point A (Figure 2–71) on the fixed vertical construction line, drag it to location B (Figure 2–71), release the mouse button, press the mouse button, drag it to location C (Figure 2–71), release the mouse button, press the mouse button, and drag it to location A (Figure 2–71). A closed spline is constructed. (Note: After the "2 Lines" sketch is dropped into the current workplane, they should be resized appropriately. Point A should be started as a point on the vertical line. When the spline is closed, the sketch will be filled automatically.)

9. Set the display to right elevation and add dimensions to the sketch in accordance with Figure 2–72.

Figure 2–71
Spline being
constructed

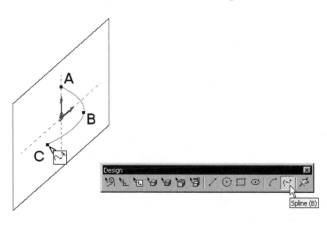

Figure 2–72
Spline constructed
and dimensioned

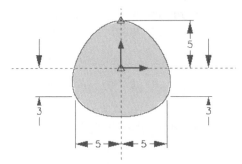

10. Set the display to trimetric.

11. Select Workplane > New Workplane.

12. In the Workplane dialog box, set selection to workplane, select Offset, and set the Offset distance to 20 mm. (See Figure 2–73.)

13. Select the lateral workplane from the browser pane and select the OK button from the Workplane dialog box. A workplane offset 20 mm from the lateral workplane is constructed.

Figure 2–73
Offset workplane
being constructed

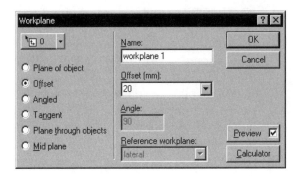

14. Construct a new sketch on the offset workplane by right-clicking on the "workplane 1" in the workplane browser.

15. With reference to Figure 2–74, construct a circle of 9.6 mm diameter.

Figure 2–74
Sketch constructed
on the workplane

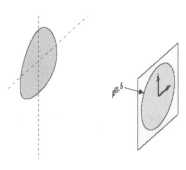

16. With reference to Figure 2–75, construct a workplane offsetting a distance of –20 mm from the lateral workplane, construct a new sketch on the offset workplane, and construct a circle of 8.4 mm diameter.

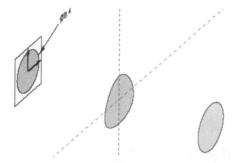

Figure 2–75
Offset sketch
constructed

Now loft through the sketch profiles to obtain a lofted solid.

1. Select Feature > Loft Through Profile.

2. Change to selection mode to Lines from the dropdown selection control in the Loft Profiles dialog box.

3. With reference to Figure 2–76, click sketch A, click sketch B, click sketch C, and select the OK button in the Loft Profiles dialog box.

A lofted solid is constructed.

Figure 2–76
Loft solid being
constructed

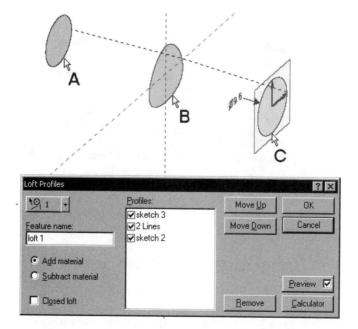

Now construct a projected solid feature to subtract material from the solid part.

1. Construct a workplane on face A indicated in Figure 2–77 and construct a new sketch on the new workplane.

2. In accordance with Figure 2–77, construct a circle of 7 mm diameter on the sketch.

3. Project the sketch to subtract through the solid part. (See Figure 2–78.)

Figure 2–77
Workplane and new
sketch constructed

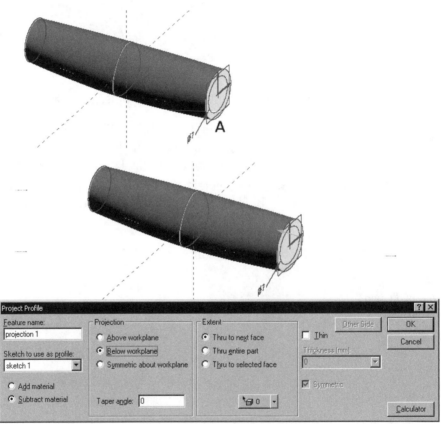

Figure 2–78 Sketch being projected to subtract through the solid part

The solid part is complete. (See Figure 2–70.) Save and close your file (file name *BallPen05.des*).

Ballpoint Pen Component and Assembly

The main body of the ballpoint pen components are complete. You will add some features to some of the components in Chapter 3, and you will put them together to form an assembly in Chapter 4.

Modification Methods

In a parametric solid modeling system, solid features can be modified both during and after construction of the solid part. Modification can be done in the many ways, as follows:

You can activate a sketch and edit the sketch to modify the solid feature constructed from the sketch. You can redefine the solid feature by changing the parameters used in making the solid feature. If a solid feature is not needed in the solid part, you can delete the feature. However, if you want to give yourself an option of retrieving an unwanted solid feature in a later stage of your design, do not delete the feature. Instead, suppress it. Sometimes, the sequence of construction of the solid features has an impact on the final form and shape of the solid part. You can reorder the operations to obtain a different result.

Edit Sketch

To change the form and shape of a sketched solid feature, you edit the sketch or the sketches that you used to construct the feature. After activating the sketch, you can delete and add sketch elements in addition to adding or removing dimension and geometric constraints. After modifying the sketch, you can propagate the change.

Activate Sketch

If there is only one sketched solid feature in a solid, you need only to select the sketch from the browser pane to highlight it and then modify the sketch in much the same way as you construct the sketch. If there are two or more sketched solid features in the solid, you can select the sketch from the browser pane, right-click, and select Activate Sketch. (See Figure 2–79.). Alternatively, you can double-click on an inactive sketch to activate it.

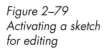

*Figure 2–79
Activating a sketch
for editing*

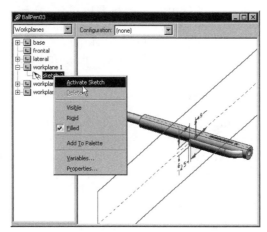

Update

After making any modification, you need to update the solid part, which can be done by selecting Feature > Update Design (Figure 2–80), selecting Update Document from the Standard toolbar, or pressing the F5 key.

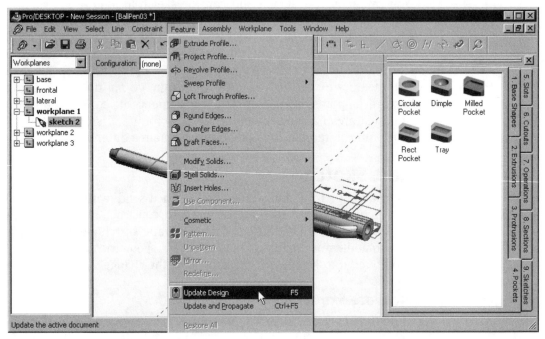

Figure 2–80 Updating a design

In essence, update calls upon the system to re-evaluate all the mathematical expressions used to define the solid part (not only the modified portion of the solid part) with the modified parameters. Because modification may cause non-congruence, invalidity, or discrepancy in the mathematical expressions of the modified feature or some other features of the solid part, update may not always be successful. If it is not, you have to edit again or undo the modification.

Update and Propagate

If there is a file or a number of files linked to a modified file, the linked file(s) will update automatically the next time you open the file(s). If a linked file is currently opened, you can cause the update to the linked file to take effect immediately by propagating the change. To update and propagate, select Feature > Update and Propagate. (See Figure 2–81.)

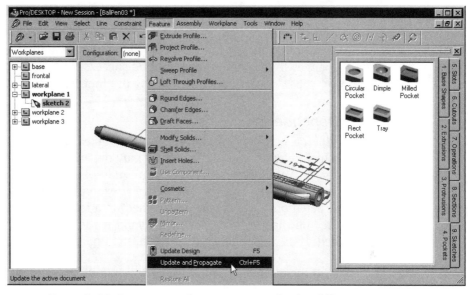

Figure 2–81 Propagating changes to opened, related files

Redefine Feature

Redefining a feature concerns modifying the parameters specified in the dialog box used to produce the feature. To activate the dialog box for modification, select the feature from the browser pane, right-click, and select Redefine. (See Figure 2–82.)

Figure 2–82
Redefining a feature

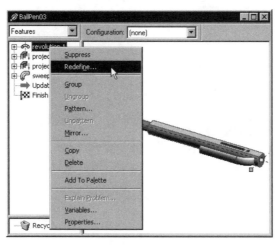

In the dialog box that follows, you can change any parameters, including the change of an Add Material operation to a Subtract Material operation or Intersect Material operation.

Update and Propagate

As with modifying a sketch, you have to update the file and, if necessary, propagate the change to the linked files that are opened. Note that changes to the feature can only be successfully updated if changes do not cause any conflict or invalidity of other features in the solid part.

Dependency

When a number of features are constructed in a solid part, they form a hierarchy in which one feature may depend on a feature constructed earlier. Validity of a certain feature may be impacted if a solid feature on which it depends is modified. For example, if a solid feature is constructed on a workplane established on a face of another solid feature, the solid feature will become invalid if the face is removed or modified. If the dependency is affected, the update may fail.

If an update fails, undo the edit and modify in some other way again.

Delete Feature

If a feature is not needed anymore, you can delete it by selecting the feature from the browser pane, right-clicking, and selecting Delete. (See Figure 2–83.) Note that deleting a feature, like redefining a feature, may cause the features depending on it to be invalid.

Figure 2–83
Deleting a solid
feature

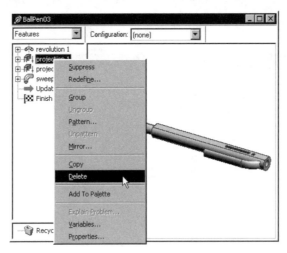

Restore

Deleted objects are placed in a Pro/DESKTOP Recycle Bin situated below the Feature browser pane. You can right-click the Recycle Bin and select Empty Recycle Bin to permanently remove the deleted object. To restore deleted objects, select them from the Recycle Bin, right-click, and select Restore. To restore all the deleted objects in a single operation, right-click the Recycle Bin and select Restore All or select Feature > Restore All from the menu bar. (See Figure 2–84.)

Figure 2–84
Restoring deleted
features from the
recycle bin

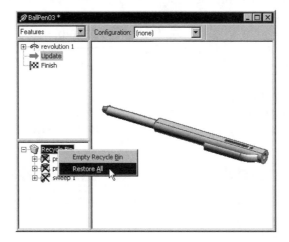

As long as the recycle bin is not emptied, you can restore deleted solid features even after saving, closing, and re-opening the file.

Suppress Feature

Features that are not needed temporarily should be suppressed rather than deleted. You can unsuppress a suppressed feature to bring it back to the solid part. To suppress a feature, select it from the browser pane, right-click, and select Suppress. (See Figure 2–85.)

Suppression, like deletion, may affect any subsequent features in the hierarchy of the solid part that depend on the suppressed features.

Figure 2–85
Suppressing feature

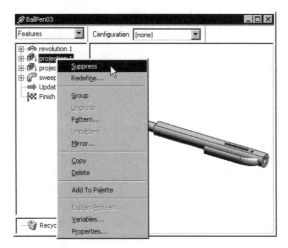

Re-Order Feature

Given the same set of sketched solid features and the same operation of combination (add material, subtract material, or intersect material), the final outcome of the solid part may be different if the order of feature construction is modified. For example, the outcome of adding a feature (A) and then subtracting a feature (B) might be different from first subtracting a feature (B) and then adding a feature (A).

To change the order of feature construction, you do not need to reconstruct the solid model; you can select a feature from the feature browser pane and drag it above another feature in the browser pane. (See Figure 2–86.)

Figure 2–86
Re-ordering features

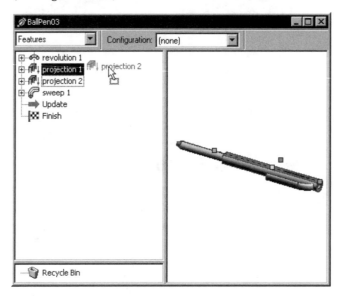

Update and Propagation

Again, you have to perform an update after re-ordering features in a solid part and propagate the change to the linked files that are opened. Note that update may not be always successful if re-ordering causes conflict and invalidity of features.

Version and Session

As a reminder here, you could save your work in a number of versions within a single file so that you can experiment with various scenarios. To save a record of the working environment of your work, you could save them in a session file. Opening the session file subsequently will retrieve all the working environments, including window sizes and locations.

Exercises

Now perform the following exercises to enhance your techniques in making sketched solid features.

Toy Plane Project

The toy plane shown in Figure 2–87 has thirteen component parts. You will construct these components in this and the next chapter and complete the entire model in Chapter 4.

Figure 2–87
Toy Plane Model

Upper Body Structure

The upper body structure shown in Figure 2–88 consists of two features, an extruded solid feature and a hole feature. You will construct the extruded solid feature here and insert the holes in the next chapter.

Figure 2–88
Upper body
structure

1. Start a new design file and set the measurement unit to millimeters.

2. With reference to Figure 2–89, construct a closed profile on the default sketch.

3. Extrude the sketch a distance of 2 mm.

4. The extruded solid feature is complete. Save and close the file (file name: Plane01.des).

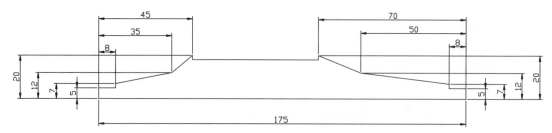

Figure 2–89 Plan view of the sketch constructed on the base workplane

Lower Body Structure

Figure 2–90 shows the lower body structure of the toy plane. It is an extruded solid.

Figure 2–90
Lower body structure

1. Start a new design file and set the measurement unit to millimeters.

2. With reference to Figure 2–91, construct a sketch on the default sketch. In the sketch, ABC is a spline.

3. Extrude the sketch a distance of 2 mm.

4. The solid part is complete. Save and close the file (file name: *Plane02.des*).

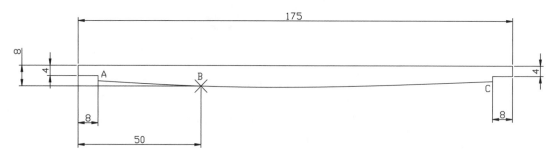

Figure 2–91 Plan view of the sketch constructed on the base workplane

Horizontal Stabilizer

Figure 2–92 shows the horizontal stabilizer of the toy plane. It is an extruded solid feature.

Figure 2–92
Horizontal stabilizer

1. Start a new design file and set the measurement unit to millimeters.

2. With reference to Figure 2–93, construct a sketch on the default sketch.

3. Add dimensions to the sketch in accordance with Figure 2–94.

4. Extrude the sketch a distance of 1 mm.

5. The solid part is complete. Save and close the file (file name: *Plane10.des*).

Figure 2–93
Plan view of the
sketch constructed
on the base
workplane

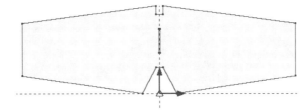

Figure 2–94
Sketch dimensioned

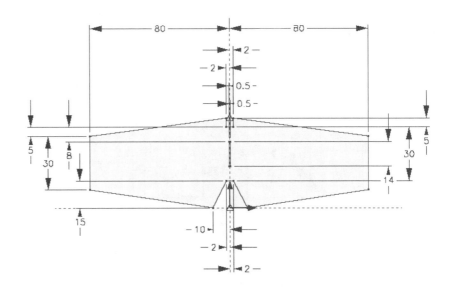

Vertical Stabilizer

Figure 2–95 shows the vertical stabilizer of the toy plane. It is also an extruded solid feature.

Figure 2–95
Vertical stabilizer

1. Start a new design file and set the measurement unit to millimeters.

2. With reference to Figure 2–96, construct a sketch on the default sketch.

3. Dimension the sketch in accordance with Figure 2–97.

4. Extrude the sketch a distance of 1 mm.

5. The solid part is complete. Save and close the file (file name: *Plane11.des*).

Figure 2–96
Plan view of the
sketch constructed
on the base
workplane

Figure 2–97
Sketch dimensioned

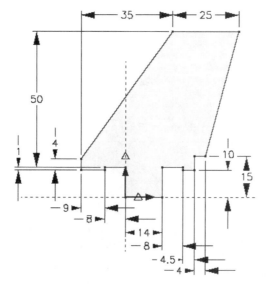

Wing

Figure 2–98 shows the wing of the toy plane. You will construct two extruded features in this chapter and insert two holes in the next chapter.

Figure 2–98
Wing

1. Start a new design file and set the measurement unit to millimeters.

2. With reference to Figure 2–99, drop the "2 Lines" sketch from the palette on the base workplane and construct two spline curves to form a closed profile. Note that A of the sketch is the origin.

3. Extrude the sketch a distance of 280 mm.

Figure 2–99
Sketch constructed
on the base
workplane

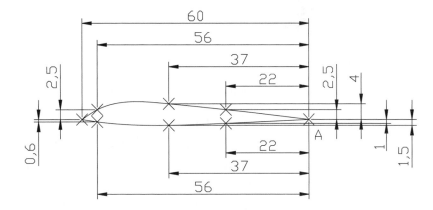

4. Construct a sketch on the lateral workplane.

5. Drop the "2 Lines" sketch from the palette on the workplane and construct a closed profile in accordance with Figure 2–100. Note that point A of the sketch is the origin.

6. Extrude the profile a distance of 60 mm and add the feature to the solid part.

7. The extruded solid features are complete. You will insert holes to complete the solid in the next chapter. Save and close the file (file name: Plane12.des).

Figure 2–100
Sketch constructed
on the lateral
workplane

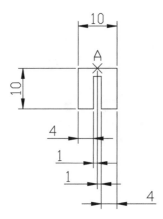

Wheel

Figure 2–101 shows the wheel of the toy plane. It has a revolved solid feature and a round feature.

Figure 2–101
Wheel

1. Start a new design file and set the measurement unit to millimeters.

2. With reference to Figure 2–102, construct a sketch on the default sketch.

3. Add dimensions to the sketch in accordance with Figure 2–103.

4. Revolve the profile about construction line A of Figure 2–103 for 360°.

5. The revolved solid feature is complete. You will add rounded edges in the next chapter. Now save and close the file (file name: *Plane08.des*).

Figure 2–102
Plan view of the
sketch constructed
on the base
workplane

Figure 2–103
Sketch dimensioned

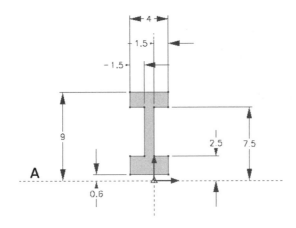

Wheel Cap

Figure 2–104 shows the wheel cap of the toy plane. It is also a revolved solid feature.

Figure 2–104
Wheel cap

1. Start a new design file and set the measurement unit to millimeters.

2. With reference to Figure 2–105, construct a sketch on the default sketch.

3. Revolve the profile about construction line A of Figure 2–105 for 360°.

4. The revolved solid feature is complete. In the next chapter, you will complete the solid part by adding rounded edges. Now save and close the file (file name: Plane09.des).

Figure 2–105
Plan view of the sketch constructed on the base workplane

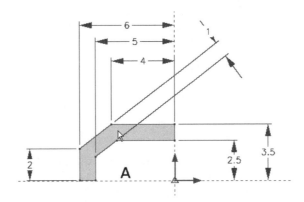

Propeller

Figure 2–106 shows the propeller of the toy plane. You will construct a lofted solid feature, a helical sweep feature, and a revolved feature in this chapter, and insert a hole feature and construct a pattern in the next chapter.

Figure 2–106
Propeller

1. Start a new design file and set the measurement unit to millimeters.

2. With reference to Figure 2–107, drop a 2-line sketch on the base workplane

3. Construct two more construction lines starting from the origin to A and to B.

4. Construct two spline curves passing through ABC and ADC.

Figure 2–107
Sketch constructed
on the base
workplane

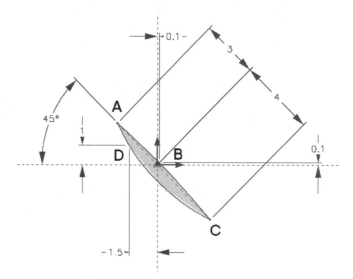

5. Construct a workplane offsetting 25 mm from the base workplane.

6. Construct a sketch on the offset workplane and drop the "2 Lines" sketch from the palette.

7. Construct two more construction lines starting from the origin to A and to B.

8. Construct two spline curves passing through ABC and ADC. (See Figure 2–108.)

Figure 2–108 Sketch constructed on a workplane offsetting 25 mm from the base workplane

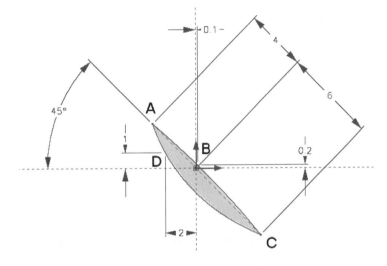

9. Construct another workplane offsetting a distance of 50 mm from the base workplane.

10. Construct a sketch on the offset workplane.

11. Set selection mode to lines, select the splines on the sketch shown in Figure 2–108, and select Line > Project to project the lines onto the current sketch.

12. With reference to Figure 2–109, construct a lofted solid feature.

Figure 2–109 Lofted solid being constructed

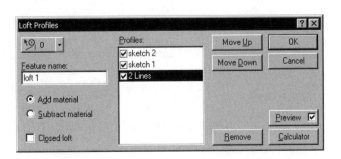

13. Construct a new workplane on the base workplane.

14. Construct a new sketch on the new workplane.

15. With reference to Figure 2–110, construct a sketch.

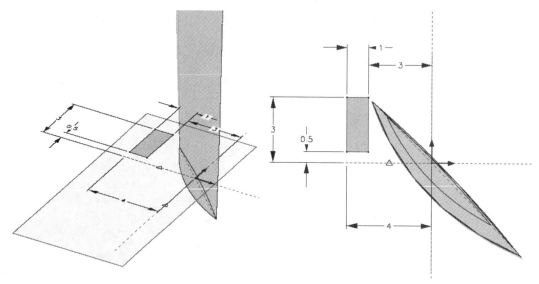

Figure 2–110 Trimetric view (left) and plan view (right) of the sketch

16. Construct a helical swept solid feature in accordance with Figure 2–111. The helix is left hand, and the pitch is 1.1 mm.

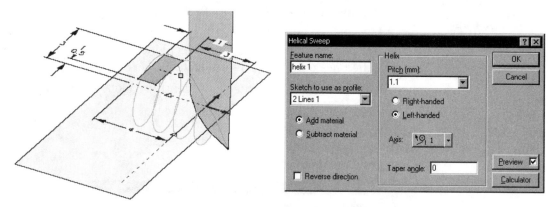

Figure 2–111 Helical swept solid being constructed

17. Construct a new workplane on the base workplane and construct a new sketch.

18. With reference to Figure 2–112, construct a sketch.

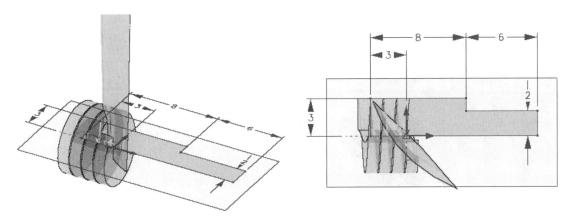

Figure 2–112 Sketch constructed on the new sketch

19. With reference to Figure 2–113, construct a revolved solid feature.

20. The lofted feature, swept feature, and revolved feature are complete.

21. Now save and close your file (file name: *Plane06.des*). You will construct a pattern of the lofted feature in the next chapter.

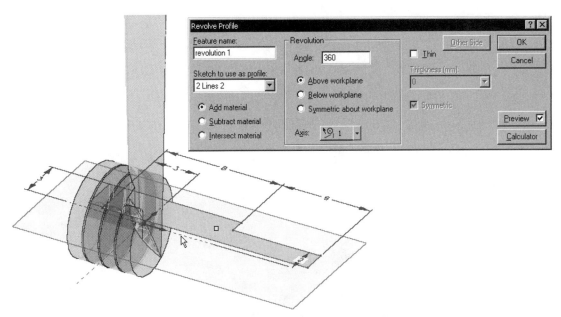

Figure 2–113 Revolved solid feature being constructed

Front Mounting

Figure 2–114 shows the front mounting of the toy plane. It has a number of features. Here in this chapter, you will construct an extruded solid feature. You will complete the model in the next chapter.

Figure 2–114
Front mounting

1. Start a new design file and set the measurement unit to millimeters.

2. Select the lateral workplane from the browser pane, right-click, and select New Sketch.

3. With reference to Figure 2–115, drop the "2 Lines" sketch and construct two closed loop profiles. Note that the vertical lines are collinear and that point A is the origin.

Figure 2–115
Right elevation
showing the sketch
constructed on the
lateral workplane

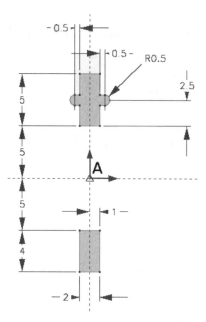

4. Set the display to trimetric and extrude the profile a distance of 10 mm above the workplane. (See Figure 2–116.)

5. The extruded solid feature is complete. Save and close the file (file name: *Plane03.des*). You will construct a shell feature and complete the model in the next chapter.

*Figure 2–116
Profile being
extruded*

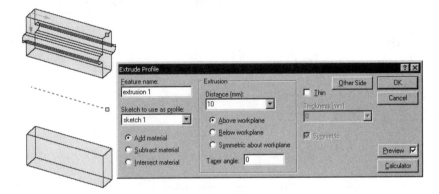

Rear Mounting

Figure 2–117 shows the rear mounting of the toy plane. It has a number of extruded solid features and a rounded edge.

*Figure 2–117
Rear mounting*

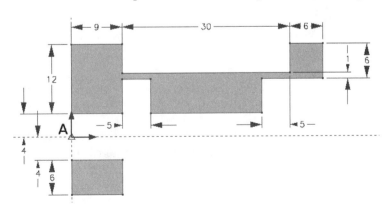

1. Start a new design file and set the measurement unit to millimeters.

2. With reference to Figure 2–118, construct a sketch on the frontal workplane. In the sketch, point A is the origin.

*Figure 2–118
Front elevation
showing the sketch
constructed on the
frontal workplane*

3. Extrude the sketch a distance of 4 mm symmetric about the workplane. (See Figure 2–119.)

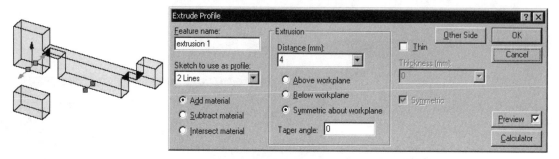

Figure 2–119 Sketch extruded symmetric about the workplane

4. Construct a new workplane on the frontal workplane.

5. Construct a new sketch on the new workplane.

6. On the sketch, construct two closed loop profiles in accordance with Figure 2–120.

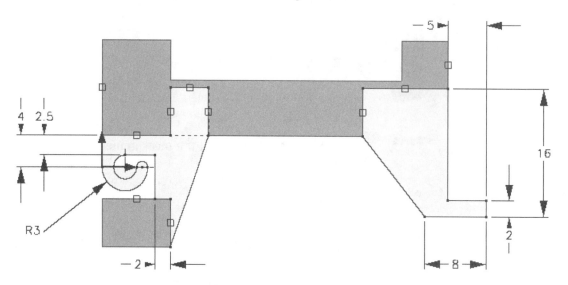

Figure 2–120 Sketch constructed on a new workplane established on the frontal workplane

7. With reference to Figure 2–121, extrude the sketch a distance of 1 mm symmetric about the workplane.

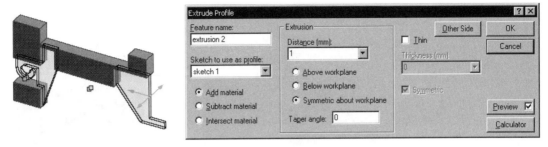

Figure 2–121 Sketch being extruded

8. Construct a new workplane on the frontal workplane.

9. Construct a new sketch on the new workplane.

10. On the sketch, construct a rectangle ABCD in accordance with Figure 2–122.

Figure 2–122
Rectangle
constructed on a
new sketch

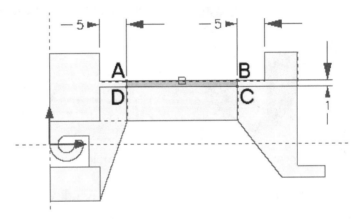

11. Extrude the sketch a distance of 12 mm symmetric about the workplane. (See Figure 2–123.)

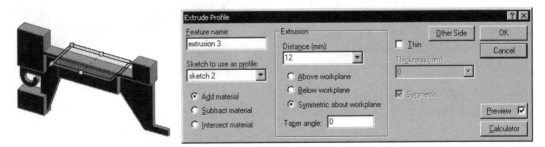

Figure 2–123 Sketch extruded symmetric about the workplane

12. Construct a sketch on face A in accordance with Figure 2–124.

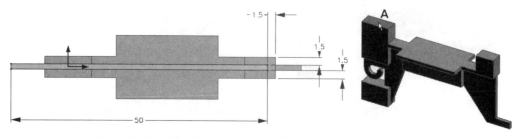

Figure 2–124 Sketch constructed on face A

13. With reference to Figure 2–125, extrude the sketch a distance of 5 mm to subtract material from the solid part.

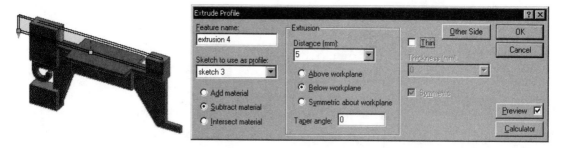

Figure 2–125 Sketch being extruded to subtract material from the solid part

14. Construct a sketch on face A indicated in Figure 2–126.

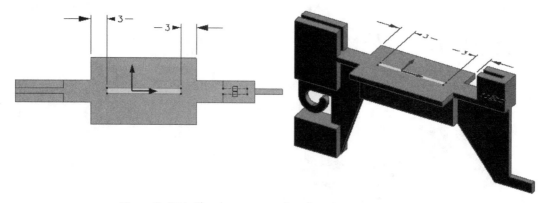

Figure 2–126 Sketch constructed on face A

15. Project the sketch to subtract through the entire solid part. (See Figure 2–127.)

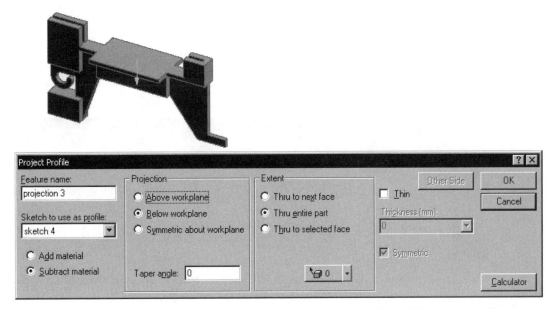

Figure 2–127 Sketch being projected to subtract through the entire solid part

16. Construct a sketch on face A indicated in Figure 2–128.

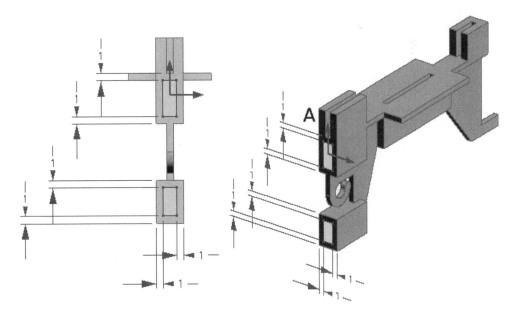

Figure 2–128 Sketch constructed

17. Extrude the sketch a distance of 8 mm to subtract material from the solid part. (See Figure 2–129.)

18. The extruded solid features are complete. Save and close your file (file name: *Plane05.des*). You will add a rounded edge in the next chapter.

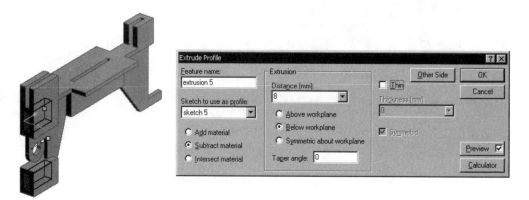

Figure 2–129 Sketch being extruded to subtract material from the solid part

Drive Shaft

The drive shaft shown in Figure 2–130 is a swept solid.

*Figure 2–130
Drive shaft*

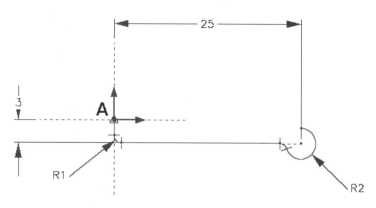

1. Start a new design file and set the measurement unit to millimeters.

2. Drop the 2-line sketch on the base workplane and construct a sketch in accordance with Figure 2–131. Point A is the origin.

*Figure 2–131
Sketch constructed
on the base
workplane*

3. Construct a sketch on the frontal workplane. (See Figure 2–132.) Point A is the origin.

Figure 2–132
Sketch constructed
on the frontal
workplane

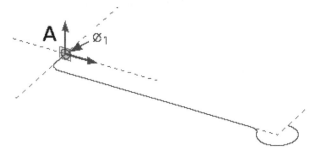

4. The solid part is complete. Save and close your file (file name: *Plane07.des*).

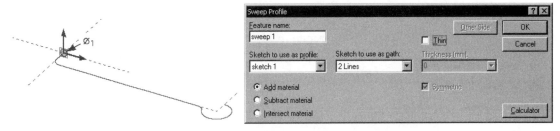

Figure 2–133 Swept solid feature being constructed

Landing Gear Main Structure

To construct the landing gear main structure shown in Figure 2–134, you will construct two swept solid features in this chapter and construct a mirror of the features in the next chapter.

Figure 2–134
Landing gear main
structure

1. Start a new design file and set the measurement unit to millimeters.

2. Construct a sketch on the base workplane. Point A of the sketch shown in Figure 2–135 is the origin.

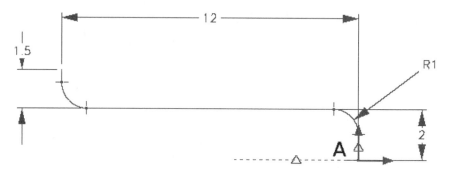

Figure 2–135 Sketch constructed on the base workplane

3. With reference to Figure 2–136, construct a sketch. The center of the circle of the sketch is the origin.

Figure 2–136
Sketch constructed
on the frontal
workplane

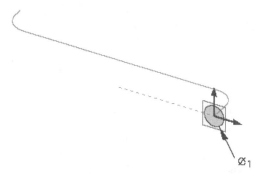

4. Using the sketch on the base workplane as the path and the sketch on the frontal workplane as the profile, construct the swept solid feature. (See Figure 2–137.)

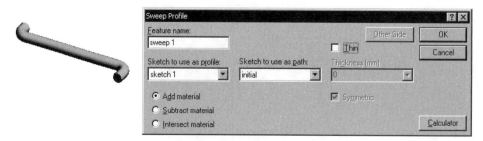

Figure 2–137 Swept solid constructed

5. With reference to Figure 2–138, construct a sketch on the lateral workplane.

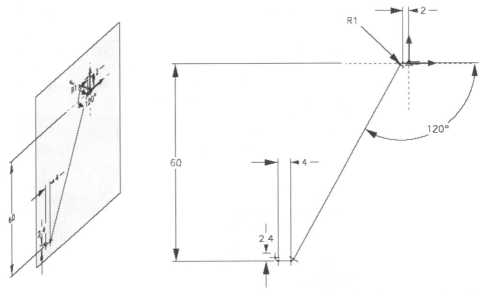

Figure 2–138 Trimetric view and right elevation of the sketch constructed on the lateral workplane

6. Construct a swept solid feature with the sketch on the lateral workplane as the path and the sketch on the frontal workplane as the profile. (See Figure 2–139.)

Figure 2–139 Second swept solid feature constructed

7. The swept solids are complete. Save and close your file (file name: Plane04.des).

Summary

Pro/DESKTOP is a feature-based parametric modeling tool for designing and constructing 3D solid models. To make a solid model, you carry out a top-down thinking process and a bottom-up construction process. In the top-down thinking process, you think about how to decompose a complex solid object into simple, unique solid features that can be constructed by using tools available from the menu. You also have to think about how to construct the features and the sequence of construction. In the bottom-up construction process, you construct the solid features one by one and combine them as you construct them.

The basic way of constructing solid features is to construct a sketch or a number of sketches to depict the cross section(s) of the feature and then treat the sketches in one of the following ways: extrude or project a profile sketch, revolve a profile sketch, sweep a profile sketch along a path sketch or along a helical path, and loft through a number of profile sketches.

A sketch requires a sketch plane, which can reside on one of the default workplanes, planar faces of the solid part, or user-defined workplanes. To properly constrain the sketch elements of a sketch, you apply dimension constraints and geometric constraints.

Because the modeling system is parametric, all the parameters used to construct the solid model are modifiable any time during or after making the solid part. Modifications can be done in several ways, including editing the sketch, redefining the features, deleting the features, suppressing the features, and re-ordering the features. After modification, you have to perform an update. Files linked to a modified solid part are updated automatically the next time you open the files. If the linked files are already opened, you can cause the update to take effect immediately by updating and propagating.

It has to be stressed that modification may cause non-congruence, invalidity, and conflict in the modified feature and any other features depending on the modified feature. Therefore, update may not always be successful. If so, perform some other modification or repair the solids that have failed as a result of the modification.

Review Questions

1. Explain what is meant by the feature-based parametric modeling approach.

2. What are the two processes involved in making a solid model? Explain them in detail.

3. List the constraints that you can apply on a sketch.

4. Delineate the ways to construct solid features from sketches.

5. State the ways to modify a solid model.

6. Distinguish between "Update" and "Update and Propagate."

CHAPTER 3

Solid Modeling II

Objectives

This chapter is a continuation of the work in Chapter 2, delineating ways to drop library objects into your design, to incorporate pre-constructed solid features, and to construct patterns and mirrors of features. After studying this chapter, you should be able to

❏   Drop library objects into your design

❏   Incorporate pre-constructed features in a design

❏   Construct patterns and mirrors of features

Overview

As you have learned in the last chapter, using the feature-based modeling approach to construct a 3D solid model consists of a top-down thinking process, to think about how to decompose the object into unique modular solid features, and a bottom-up construction process, to make individual solid features one by one, combining them in a single solid part as you construct them. To reduce design lead time, you can use library objects in your design. Besides making solid features by sketching, you can incorporate commonly used solid features by selecting them from the menu and specifying their parameters. If there are repeated features in your design, you can construct feature patterns and mirror features.

Palette—Standard Object Library

A quick way to make a solid model is to make use of the objects available from the palette, which is a library of standard sketches, features, and components. Figure 3–1 shows the palette with the Base Shapes tab displayed on the top.

Figure 3–1
Palette—Base
Shapes tab

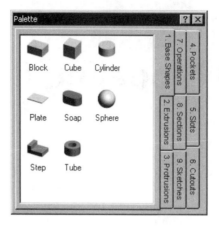

If the palette is not displayed, check the Palette check box from the Tools menu to open it. To close the palette, either select the [x] mark at the upper left corner of the palette or clear the Palette check box from the Tools menu.

Dropping Objects from the Palette

Objects from the palette can be dropped into your design by double-clicking the object or selecting and dragging the object to the active design window. Depending on the dropping method setting of the object, prior to double-clicking or selecting and dragging, you have to select an edge, a face, or a workplane for placing the object. For example, you have to select an edge prior to dropping the Round Edge object, and to select a face prior to using an extrusion object. To set the pre-selected placement object type, right-click the object's icon from the palette and select Drop onto Edge, Drop onto Face, or Drop onto Active Workplane. (See Figure 3–2.)

Figure 3–2
Right-click menu of
objects in the palette

Modification of Parameters

You can modify the parameters of a sketch or a feature immediately after dropping it into your design by right-clicking the object's icon in the palette and checking Redefine On Drop. Immediate modification can be done in several ways. If a dialog box is displayed, you can modify the parameters in the dialog box. If handles of the object are displayed, you can manipulate the handles to change its size and orientation. (See Figure 3–3.) If a sketch is inserted, you can manipulate the sketch's constraints. Naturally, you can use the methods explained in Chapter 2 to modify various aspects of the object at any time later.

Figure 3–3
Dragging the handles of a dropped block from the base shapes tab of the palette

Default Palette

Among the three kinds of standard library objects, the default palette has only sketch objects and feature objects. (The use of component objects will be dealt with in Chapter 5.) They are saved in nine subfolders of the Palette folder in the Pro/DESKTOP Program folder, displayed in nine tabs of the palette: Base Shapes, Extrusions, Protrusions, Pockets, Slots, Cutouts, Operations, Sections, and Sketches. In accordance with the nature of the objects, they can be classified into five categories, as follows.

Base Shapes, Extrusions, and Protrusions

The first category of library objects adds material to your design. The objects are stored in the Base Shapes tab, the Extrusions tab, and the Protrusions tab. Objects here are sketched solid features. Before picking them from the palette, you have to select a face or a workplane for placement of the sketch defining the feature.

The Base Shapes tab consists of the following objects: block, cube, cylinder, plate, soap, sphere, step, and tube. They are basic building blocks for solid modeling. (See Figure 3–1.) The Extrusions tab is a library of extruded solid models suitable for designing structural parts. It has the following objects: box section, channel, equal angle, equal tee, I beam, rectangular section, tube, unequal angle, and unequal tee.

(See Figure 3–4.) The Protrusions tab has the following objects: boss, bump, frustum, rectangular pad, roll, round boss, round frustum, square pad, and tube. These objects are suitable for detailing a solid model. (See Figure 3–5.)

Figure 3–4
Extrusions tab

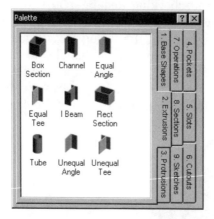

Figure 3–5
Protrusions tab

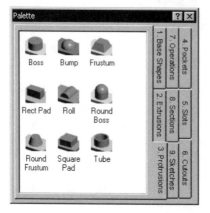

Pockets, Slots, and Cutouts

The second category of library objects subtracts material from your design. These objects are also sketched solid features, and they are saved in the Pockets, Slots, and Cutouts tabs of the palette. Again, you need to pre-select a face or a workplane for placement of the defining sketch of the features. Because you cannot subtract material from nothing, any individual object from these tabs cannot be used as the first object of your design.

The Pockets tab consists of the following objects: circular pocket, dimple, milled pocket, rectangular pocket, and tray. As the name of the tab implies, these objects are mainly used for cutting out pockets of various shapes. (See Figure 3–6.) Objects in the Slot tab cut slots of various shapes in your design, including circular slot, round slot, simple slot, t-slot, and v-slot. (See Figure 3–7.) The Cutouts tab provides objects for cutting through your design with the shape of a hole, keyhole, slot, and slotted hole. (See Figure 3–8.)

Figure 3–6
Pockets tab

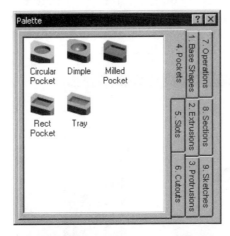

Figure 3–7
Slots tab

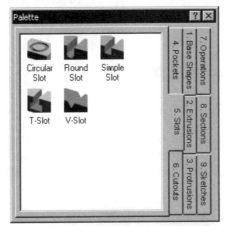

Figure 3–8
Cutouts tab

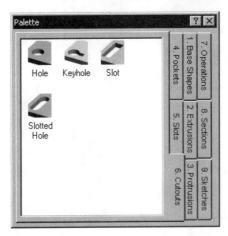

Operations

The objects delineated in this tab are actually replicating some of the pre-constructed solid features. The chamfer edge and round edge objects require an edge or a number of edges to be pre-selected, and the chamfer face, round face, and shell require a face or a number of faces to be pre-selected. (See Figure 3–9.)

Figure 3–9
Operations tab

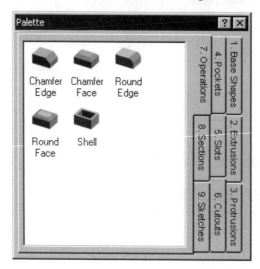

Sections

The Sections tab has three objects: curved section, stepped section, and straight section. Objects here are quite special, in that they cut away approximately half of your design with different section planes. (See Figure 3–10.)

Figure 3–10
Sections tab

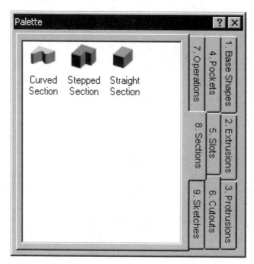

Sketches

The Sketch tab has nine objects: 2 lines, 4 lines, 5 star, hexagon, octagon, pentagon, rectangle, square, and triangle. They quickly provide a sketch of various shapes on the pre-selected face or workplane. (See Figure 3–11.)

Figure 3–11
Sketches tab

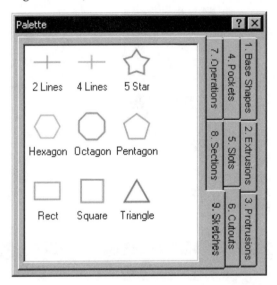

Customizing the Palette

The palette is customizable, in that you can add new objects and tabs to it as well as delete existing objects and tabs from it.

Adding Palette Objects and Palette Tabs

A way to reduce design lead time is to save sketches, features, and components already constructed in the palette as standard library objects and re-use them in new designs by retrieving them from the library.

Before putting a sketch or a feature in a tab of the palette, you have to select the tab to bring it to the top of the palette. To add a sketch to the active tab of the palette, set the browser pane to display workplanes, right-click the sketch from the browser pane, and select Add to Palette. (See Figure 3–12.) To add a feature to the active tab, set the browser pane to display features, right-click the feature from the browser pane, and select Add to Palette. (See Figure 3–13.) Once a sketch or a feature is added to a tab of the palette, a file representing the sketch or the feature will be saved in the subfolder of the palette folder. The file extension for a sketch is *.ske* and for a feature, *.fea*.

Figure 3–12
Adding selected
sketch to the palette

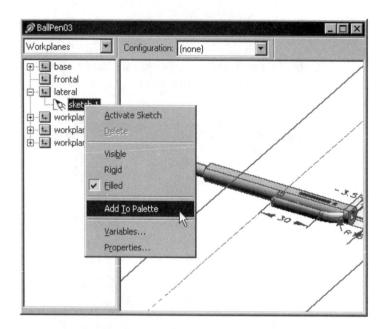

Figure 3–13
Adding selected
feature to the palette

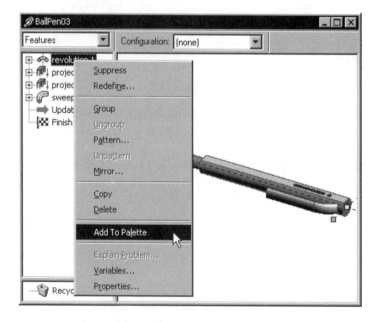

To add a component to a tab of the palette, copy the component's design file to the subfolder of the Palette folder of the Pro/DESKTOP folder. An icon representing the component will appear in the selected tab of the palette the next time you open the palette. If you want to change the icon, right-click the icon from the palette, select Set Icon (Figure 3–14), and select a bitmap file.

Figure 3–14
Setting icon image

To help organize objects in a palette, you can add or delete tabs in a palette. To add a tab in a palette, right-click an empty space in one of the tabs of the palette, and select New Folder. (See Figure 3–15.) In the New Folder dialog box, specify a name for the new tab.

Figure 3–15
Adding a new folder
in the palette

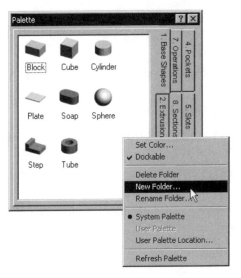

Deleting Palette Objects and Palette Tabs

In addition to adding objects and tabs to the palette, you can delete unwanted objects and tabs from it. To delete an object, right-click it on the palette tab and select Delete. To delete a tab, select a tab to bring it to the top, right-click an empty area, and select Delete Folder. Naturally, you can manipulate objects through the use of Windows Explorer. However, it is strongly recommended that the default palette folder and the objects in the folders not be deleted.

Incorporating Pre-Constructed Features

The second kind of solid feature that you can use to compose a solid model is a pre-constructed solid feature available from the menu. Pre-constructed solid features, including Round Edges, Chamfer Edges, Draft Faces, Shell Solids, Insert Holes, and Cosmetic Screw Thread, are features commonly used in most solid models.

Except for insert hole, in which you need to construct a sketch to depict the location of centers of the holes, construction of the other kinds of pre-constructed features does not need any sketch.

Round Edges

It is a common design practice to round off the edges of a 3D solid model to fulfill aesthetic or functional requirements. Using the round edges command, you specify a constant radius or a variable radius for the selected edges. Except for aesthetic reasons, rounding edges should be done at the final stage of modeling.

Now start a new design file, drop a block into the design file, and construct round edges.

1. Select File > New > Design or select Design from the Standard toolbar.
2. Set the unit of measurement to mm.
3. Double-click the Block icon of the Base Shapes tab of the palette.
4. Click an empty space in the design pane to de-select the dropped block.
5. Select Feature > Round Edges or select Round Edges from the Features toolbar.
6. Set selection mode to Edges and select edge AB indicated in Figure 3–16.
7. In the Round Edges dialog box, select Constant radius, if it is not already selected, set the radius to 12 mm, and select the OK button.

A round edge with a constant radius of 12 mm is constructed.

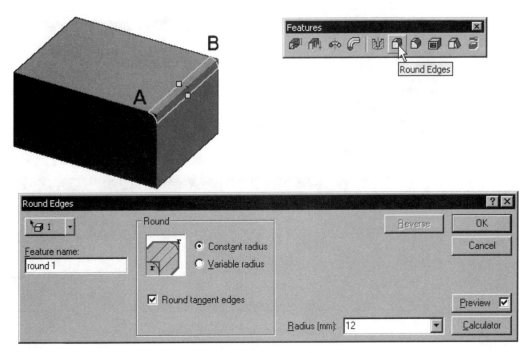

Figure 3–16 Constant radius round edge being constructed

8. Select Feature > Round Edges or select Round Edges from the Features toolbar.

9. Select edge AB indicated in Figure 3–17.

10. In the Round Edges dialog box, select Variable radius, double-click the first row under the % Length column and set the position to 0, and double-click the first row under the Radius column and set the radius to 12 mm. The radius at the starting endpoint of the selected edge is set. (See Figure 3–17.)

11. Double-click the second row of the % Length column and set the position to 30.

12. Double-click the second row of the Radius column and set the radius to 20. The radius of the edge at a location equal to 30% of the length from the starting end point is set to 20 mm.

13. Repeat steps 11 and 12 to set the radius at 100% of the edge length to 25 mm.

14. Select the OK button.

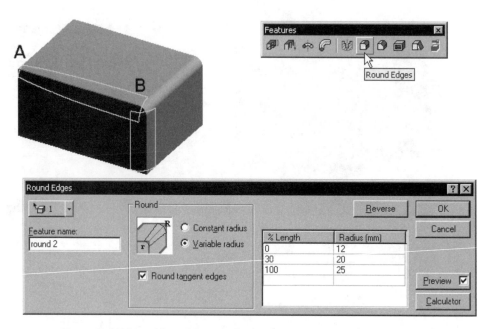

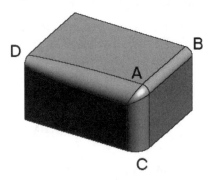

Figure 3–17 Variable radius round edge being constructed

A variable radius round edge is constructed. Because edge AB is already rounded when you round edge AD, edge AC is rounded as well, using the radius at the end point of edge AD as its radius. (See Figure 3–18.)

Figure 3–18
Round edges
constructed

The model is complete. Save and close your file (file name: *RoundEdge.des*).

Chamfer Edges

Another way to treat the sharp edges of a solid model is to chamfer them. Chamfering concerns beveling the edges by setting back the faces on both sides of the selected edges. You can apply an equal setback distance, an unequal setback distance, or a setback distance together with a bevel angle to specify the size and shape of the beveled edge. Like rounding the edges, chamfering should be done at the final stage of modeling.

Now start a new design file, drop a block into the design file, and construct chamfer edges.

1. Select File > New > Design or select Design from the Standard toolbar.

2. Set the unit of measurement to mm.

3. Double-click the Block icon of the Base Shapes tab of the palette.

4. Click an empty space in the design pane to de-select the dropped block.

5. Select Feature > Chamfer Edges or select Chamfer Edges from the Features toolbar.

6. Set selection mode to Edges and select edge AB indicated in Figure 3–19.

7. In the Chamfer Edges dialog box, select Equal setback, set setback to 25 mm, and select the OK button.

The selected edge is chamfered. Save and close your file (file name: *ChamferEdge.des*).

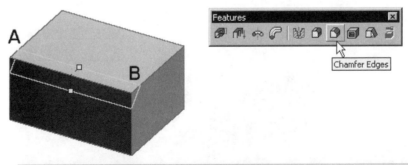

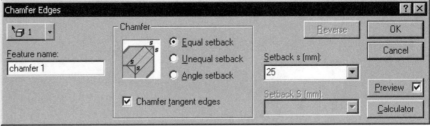

Figure 3–19 Edge being chamfered

Draft Faces

To facilitate removal of a component from a mould for making the component, you apply a draft angle to vertical faces by using the draft faces command.

Now start a new design file, drop a block into the design file, and construct draft faces.

1. Select File > New > Design or select Design from the Standard toolbar.

2. Set the unit of measurement to mm.

3. Double-click the Block icon of the Base Shapes tab of the palette.

4. Click an empty space in the design pane to de-select the dropped block.

5. Select Feature > Draft Faces or select Draft Faces from the Features toolbar.

6. Select face A indicated in Figure 3–20.

7. In the Draft Faces dialog box, set the draft angle to 5, select the base workplane as the neutral plane, and select the OK button.

A draft angle of 5° with reference to the base workplane is applied to face A. Save and close your file (file name: *DraftFace.des*).

Figure 3–20
Draft face being
applied

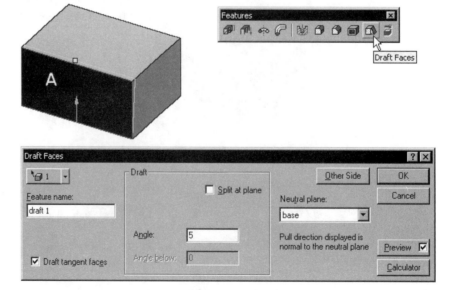

Shell Solids

The shell solids command makes a solid part hollow. To construct an open shell solid, you remove a face or a number of faces from the original solid part while shelling.

Now start a new design file, drop a block into the design file, and shell the solid part.

1. Select File > New > Design or select Design from the Standard toolbar.

2. Set the unit of measurement to mm.

3. Double-click the Block icon on the Base Shapes tab of the palette.

4. Click an empty space in the design pane to de-select the dropped block.

5. Select Feature > Shell Solids or select Shell Solids from the Features toolbar.

6. Select face A indicated in Figure 3–21.

7. In the Shell Solids dialog box, set the offset value to 4 mm in the Offset tab, select Inside, and select the OK button.

The solid is shelled. (See Figure 3–22.) Save and close your file (file name: *ShellSolid.des*).

Figure 3–21
Solid being shelled

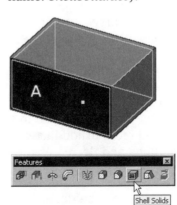

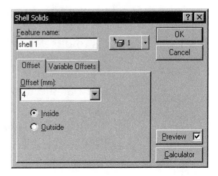

Figure 3–22
Solid part shelled
and a face removed

Offsetting and Subtracting

Although shell solids is a single command, the shelling process involves two operations: offsetting faces and subtracting material. The process of offsetting (inside or outside) the faces of a solid part produces another solid part of similar shape but smaller (offsetting inside) or larger (offsetting outside) in size. If offsetting is made inside the original solid part, the solid formed from offsetting faces is subtracted from the original solid part. If offsetting is done outside the original solid part, the original solid is subtracted from the solid formed from offsetting faces. In both cases, a hollow solid is constructed.

Figure 3–23
Inside offset (left)
and outside offset
(right)

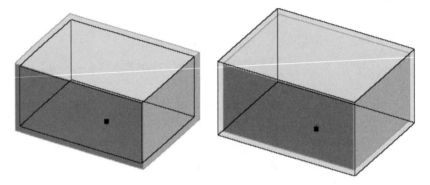

Failure to Shell

It is very important to note that shelling solid parts with curved faces (a rounded edge is a kind of curved face) may not be always successful, because the operation will cease if one of the faces of the solid part fails to offset.

Offsetting curved faces outside concerns increasing the radius of curvature at the convex portions and decreasing the radius of curvature at the concave portions for a distance equal to the offsetting distance. Similarly, offsetting curved faces inside increases the radius of curvature at the concave portions and decreases the radius of curvature at the convex portions.

Mathematically, offsetting cannot be done if the offsetting distance is larger than the smallest radius of curvature at the concave portion of a curved face in an outside offsetting operation or at the convex portion of the curved faces in an inside offsetting operation.

If shelling fails, increase the radius of curvature at the problematic area of the solid part or reduce the offsetting distance.

Figure 3–24 shows a failure to shell a solid part with a variable radius round edge.

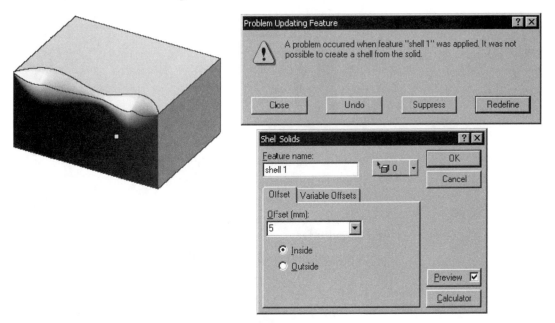

Figure 3–24 Failure to shell

Cosmetic Screw Thread

A screw thread is a helical groove cut on a cylindrical hole or shaft. To construct a solid model of a screw thread, you can first construct a cylindrical solid feature, construct a helical swept solid feature, and then subtract the helix swept solid feature from the cylindrical solid feature. If the screw thread to be constructed is a standard engineering feature and it is unnecessary to spend so much time and effort in making the swept feature, you can incorporate cosmetic screw threads instead. A cosmetic screw thread is a helical curve put on the cylindrical hole or shaft to symbolize the thread.

Now start a new design file, drop a cylinder into the design file, and add a cosmetic screw thread.

1. Select File > New > Design or select Design from the Standard toolbar.

2. Set the unit of measurement to mm.

3. Double-click the Cylinder icon on the Base Shapes tab of the palette to drop it onto the default base workplane.

4. Add a dimension of R20 to the sketch of the cylinder. (See Figure 3–25.)

*Figure 3–25
Dimension added
to the dropped
cylinder*

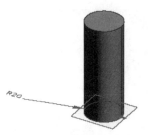

5. Select face A indicated in Figure 3–26.

6. Select Feature > Cosmetic > Screw Thread.

7. In the Thread Feature dialog box, select the OK button. A cosmetic screw thread is applied to the entire length of the cylinder.

8. Select the screw thread from the browser pane; the cosmetic screw thread is highlighted.

The solid part is complete. Save and close your file (file name: *CosmeticThread.des*).

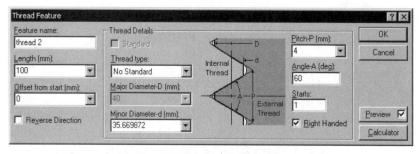

Figure 3–26 Cosmetic screw thread being constructed

*Figure 3–27
Cosmetic screw
thread selected and
highlighted*

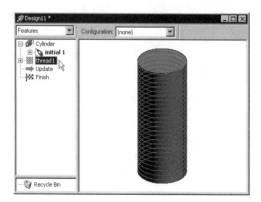

Insert Hole

Hole features can be incorporated into a solid part by sketching a number of circles to depict the center location of the holes. The diameter of the circles is unimportant because only the center positions of the circles are taken into account when hole feature is inserted.

A hole incorporated this way can be a simple hole, a tapered hole, a counterbored hole, a countersink hole, or a counterdrilled hole. In addition, you can incorporate cosmetic threads on the hole.

Now start a new design file, drop a block into the design file, and construct a sketch.

1. Select File > New > Design or select Design from the Standard toolbar.

2. Set the unit of measurement to mm.

3. Double-click the Block icon on the Base Shapes tab of the palette.

4. Select face A indicated in Figure 3–28, right-click, and select New Sketch. In the New Sketch dialog box, select the OK button. As you can see from the browser pane, a workplane is constructed on the selected face, and a sketch is established on the workplane.

Figure 3–28
Sketch being
established

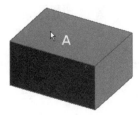

5. With reference to Figure 3–29, construct two circles and add dimensions to constrain the center position of the circles. As mentioned earlier, the circle's diameter is unimportant and therefore not specified.

Figure 3–29
Sketch constructed

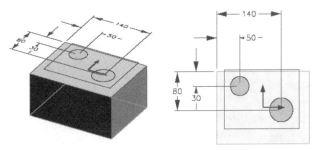

6. Select Feature > Insert Holes or select the Insert Holes button from the Feature toolbar.

7. In the Insert Hole dialog box, select Counterbore to insert counterbore holes. (See Figure 3–30.)

8. Set C'bore diameter to 30 mm, C'bore depth to 15 mm, Depth to 50mm, and diameter to 15 mm.

9. Select the OK button.

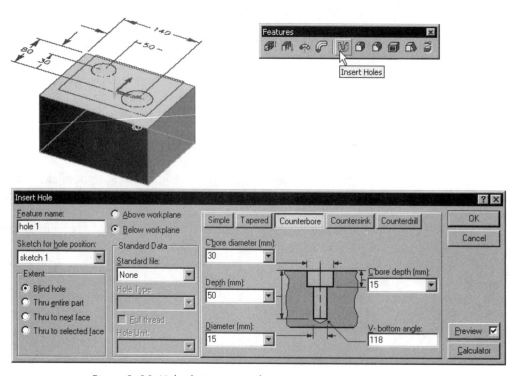

Figure 3–30 Holes being inserted

The holes are inserted. Set the view mode to transparent by using the shortcut key F10. (See Figure 3–31.) Save and close your file (file name: *InsertHole.des*).

Figure 3–31
Holes inserted

Repeated Features

Quite often, features are repeated in a solid model for various reasons. If you want to construct a set of similar features along a straight line path or a circular path, you can construct a pattern of features instead of repeatedly building the features one by one. If a component is symmetric about a face or a workplane, you can repeat the features of the component in the form of mirror copies.

Pattern

A pattern duplicates selected features in one or two directions, which can be linear path and/or circular path, resulting in three different kinds of patterns: pattern in the form of a parallelogram, circular pattern in a single ring, and circular pattern in multiple rings, as shown in Figure 3–32.

Figure 3–32
Dimples on a block
(left) and various
kinds of patterns of
the dimples (right)

Pattern Construction

Constructing a pattern is done by first selecting the features to be duplicated, right-clicking, and selecting Pattern or selecting Feature > Pattern. In the Pattern Feature dialog box that follows (shown in Figure 3–33), specify one or two directions, the number of instances along the pattern direction, and the spacing of instances. Check the Ignore member failures check box to suppress any instances that cannot be constructed.

Figure 3–33
Pattern Feature
dialog box

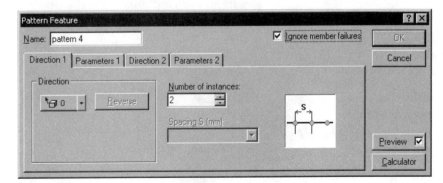

Direction, Instances, and Spacing

As we have mentioned, pattern direction can be linear or circular. To specify a direction, you select an edge or a sketch line (straight line or circular arc). If no suitable edges or sketches are available, you have to construct a sketch with straight lines and/or circular arc to depict the pattern directions beforehand.

The number of instances specified in the Pattern Feature dialog box includes the source. Therefore, if you state an instance number of 3, you will construct two instances in addition to the source. Depending on the nature of the direction path, which can be linear or circular, you specify the spacing between adjacent instances in terms of a linear distance or an angle.

Pattern with Linear Directions

In a pattern, you can specify just one direction path or two direction paths. If you specify one linear path, feature instances will repeat along the path in a linear row. If you specify two linear paths, you obtain a pattern in a parallelogram. It must be stressed that the two linear paths need not be perpendicular to each other.

Now start a new design file, construct a block and a dimple, and make a pattern of the dimple.

1. Select File > New > Design or select Design from the Standard toolbar.

2. Set the unit of measurement to mm.

3. Double-click the Block icon from the Base Shapes tab of the palette to drop it onto the base workplane.

4. Select face A (Figure 3–34) of the dropped block.

5. Double-click the Dimple icon on the Pockets tab of the palette to drop a dimple onto the upper face of the block. (See Figure 3–34.)

Figure 3–34
Block dropped on the base workplane and a dimple dropped onto the top face of the block

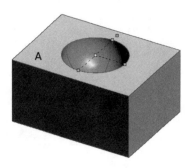

6. Select and drag the handle of the dropped dimple to resize it to 14x28.

7. Select the dimple feature and drag it to a new location as indicated in Figure 3–35.

Figure 3–35
Dimple resized and
dragged to new
location

8. Select the dimple feature, right-click, and select Pattern.

9. Select edge A indicated in Figure 3–36.

10. Select the Direction 1 tab of the Pattern Feature dialog box, if it is not already selected, and set the number of instances to 4 and spacing to 40.

11. Select the Direction 2 tab of the Pattern Feature dialog box and set the number of instances to 3 and spacing to 35.

12. Select edge B indicated in Figure 3–36. Make sure that the arrow points in the right direction. If not, reverse it.

13. Select the OK button.

A pattern in two linear directions is constructed. Save and close your file (file name: *PatternRectangular.des*).

Figure 3–36
Pattern in the form
of a parallelogram
being constructed

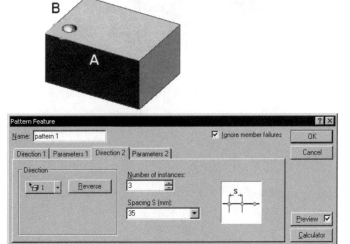

Circular Pattern

If you specify a circular path, you obtain a circular pattern. If you specify a circular path as the first pattern direction together with a linear path as the second pattern direction, you obtain a circular pattern in multiple rings.

Now start a new design file, construct a block and a dimple, and make a multiple-ring circular pattern of the dimple.

1. Select File > New > Design or select Design from the Standard toolbar.

2. Set the unit of measurement to mm.

3. Double-click the block icon on the Base Shapes tab of the palette to drop it onto the base workplane.

4. Select face A indicated in Figure 3–37, right-click, select New Sketch, and select the OK button from the New Sketch dialog box.

5. Construct a circle. The diameter is unimportant because only its center will be used for making a circular pattern.

Figure 3–37
Circle constructed
on a sketch
established on
the top face of a
dropped block

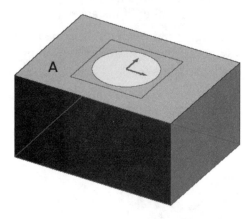

6. Select face A indicated in Figure 3–38.

7. Double-click the Dimple icon on the Pockets tab of the palette to drop a dimple onto the upper face of the block

8. Select and drag the handle of the dropped dimple to resize it to 14x28.

9. Select the dimple feature and drag it to a new location as indicated in Figure 3–38.

Figure 3–38
Dimple feature
dropped, resized,
and repositioned

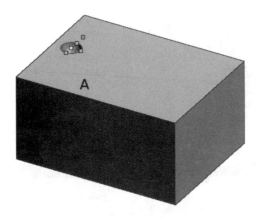

10. Select the dimple feature, right-click, and select Pattern.

11. Select edge A indicated in Figure 3–39.

12. Set the number of instances to 2 and the spacing to 40.

Figure 3–39
Linear direction
specified

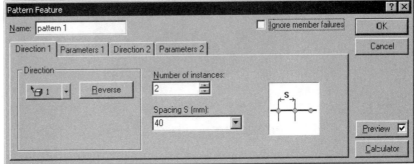

13. Select the Direction 2 tab of the Pattern Feature dialog box.

14. Set selection mode to Line and the select the circle.

15. Set the number of instances to 6 and the angle to 60.

16. Select the OK button.

*Figure 3–40
Circular path
selected*

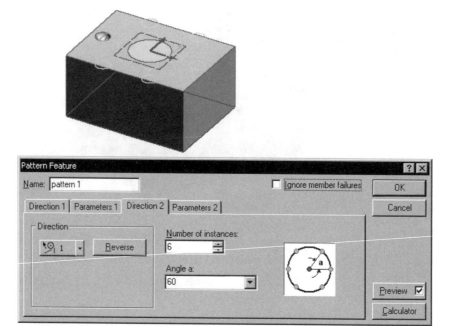

*Figure 3–40
Circular path
selected*

A multiple ring circular pattern is constructed. (See Figure 3–41.) Save and close your file (file name: *PatternCircular.des*).

*Figure 3–41
Circular pattern
constructed*

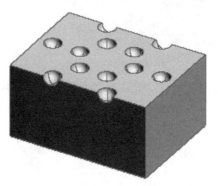

Parameters

The repeated features in a pattern are called instances. If you re-define the source, the instances will follow the change. With regard to size and shape, the instances are basically the same as the source from which the instances are duplicated. However, you can apply an incremental size change to the instances by setting the parameters of the features or the sketches of the features.

Among the four tabs in the Pattern Feature dialog box, the Parameters 1 and Parameters 2 tabs are devoted to incremental change in feature parameters and sketch parameters of the instances in Direction 1 and Direction 2 respectively.

Feature Parameters are parameters used in making the feature; these parameters are automatically included in the list. Sketch Parameters are parameters of the sketch of the feature in the pattern. Sketch parameters are not included in the list until you select them from the sketch while the Pattern Feature dialog box is opened. If you click the cell adjacent to the name of the parameters and specify the incremental value, parameters of successive instances of a pattern will increase accordingly.

Now re-define a pattern by incorporating incremental change in feature parameters and sketch parameters.

1. Open the file *PatternRectangular.des*

2. Activate the sketch of the dimple feature and change the diameter of the dimple to 24 mm. (See Figure 3–42.)

Figure 3–42 Diameter of dimple set

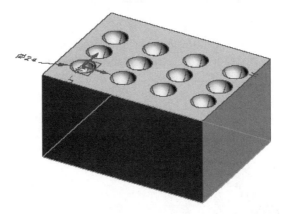

3. Set the browser pane to display the features of the solid part.

4. Select the pattern feature from the browser pane, right-click, and select Redefine.

5. Select Parameter 1 tab from the Pattern Feature dialog box.

6. Select the dimension indicated in Figure 3–43 and set the incremental change of the dimension to 5. The dimple's diameter will increase by 5 mm for each instance in the first direction.

7. Set the incremental change of the dimple/angle to –30. The dimple's rotation angle will change by –30 degrees for each instance in the first direction.

Figure 3–43
Feature parameter
and sketch
parameter of the
first pattern direction
changed

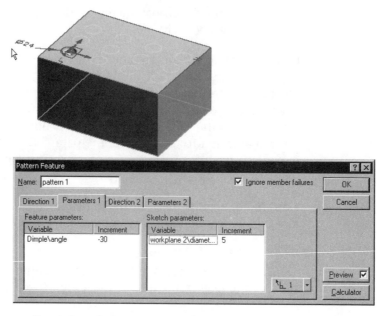

8. Select the Parameter 2 tab.

9. Select the dimension indicated in Figure 3–44 and set the incremental change in dimension to 10.

10. Select the OK button.

The parameters of the instances of the pattern increase in both directions of the pattern. (See Figure 3–45.) Save and close your file.

Figure 3–44
Feature parameter
and sketch
parameter of the
second pattern
direction changed

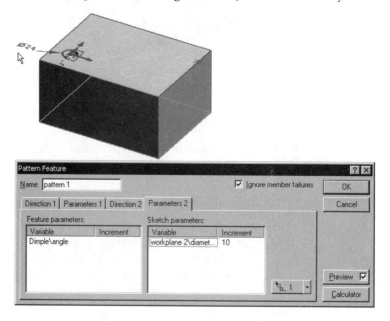

*Figure 3–45
Effect of changed
parametersPattern's
parameter changed*

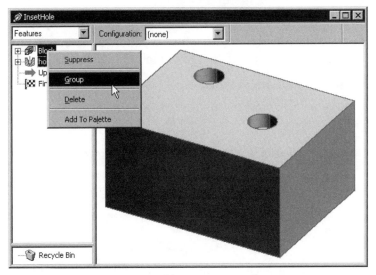

Figure 3–46 Grouping a set of features

Group

Grouping is a means of managing a set of features so that you can apply a command to a group collectively. To construct a group, select a set of features from the browser pane, right-click, and select Group. (See Figure 3–46.) To remove a group, select the group from the browser pane, right-click, and select Ungroup. (See Figure 3–47.)

*Figure 3–47
Ungrouping a group
of features*

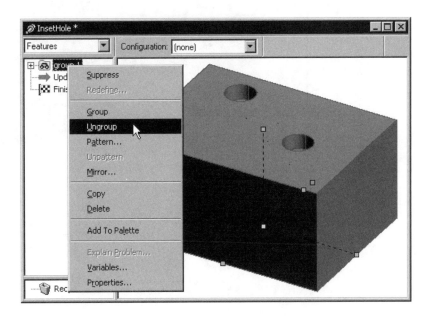

When a pattern or a mirror copy is being produced from a selected feature or a number of selected features, the selected source feature(s) are put into a group automatically.

Unpattern and Ungroup

By default, instances are linked to the source. To change the parameters of an instance individually independent of the source, you have to first unpattern the pattern by selecting the pattern from the browser pane and selecting Unpattern.

Now perform the following steps to unpattern a pattern feature, ungroup the grouped instances, and modify an individual unlinked instance.

1. Open the file *PatternCircular.des*.

2. Select the pattern feature from the browser pane, right-click, and select Unpattern. (See Figure 3–48.)

Figure 3–48
Pattern feature being unpattered

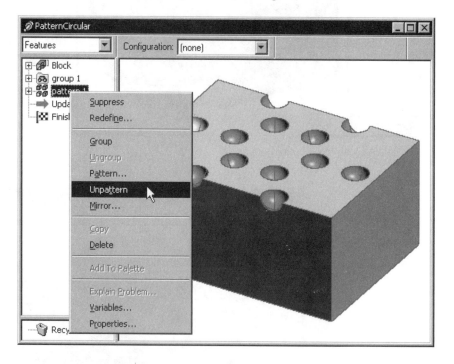

3. Select Feature > Update Design.

4. Select the groups from the browser pane one by one, right-click, and select Ungroup. (See Figure 3–49.)

*Figure 3–49
Groups being
ungrouped*

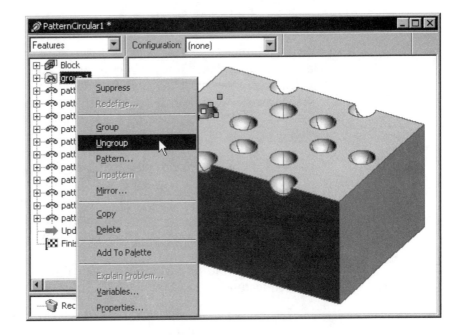

5. With reference to Figure 3–50, modify an unlinked instance.

*Figure 3–50
Unlinked instance
being modified*

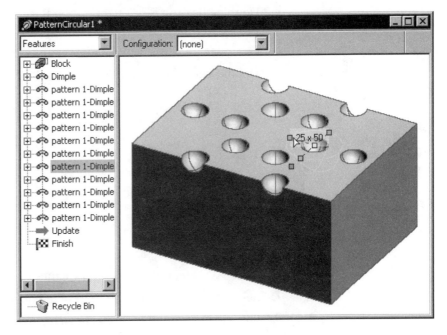

The pattern is modified. Save and close your file.

Mirror

Mirror produces a mirror copy of a set of selected features about a face or a workplane. To construct a mirror copy of a feature or a set of features, you select the features, right-click, and select Mirror or

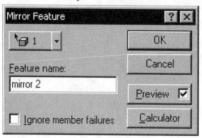

select Feature > Mirror. In the Mirror Feature dialog box that follows (shown in Figure 3–51), set the selection object type, face or workplane, select a mirror plane, and select the OK button. To suppress failure mirror objects, check the Ignore member failures check box.

Figure 3–51 Mirror Feature dialog box

Now perform the following steps to construct a mirror copy of a selected feature.

1. Select File > New > Design or select Design from the Standard toolbar.

2. Set the unit of measurement to mm.

3. Double-click the Block icon on the Base Shapes tab of the palette to drop it onto the base workplane.

4. Select face A of the block indicated in Figure 3–51.

5. Double-click the Round Frustum on the Protrusions tab of the palette.

Figure 3–52
Round frustum
dropped onto a face
of a block

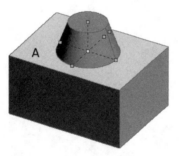

6. Select Workplane > New Workplane.

7. Select Mid plane from the Workplane dialog box.

8. Set selection mode to faces and select faces A and B indicated in Figure 3–53.

9. Select the OK button.

A workplane is constructed.

Figure 3–53
Workplane being
constructed

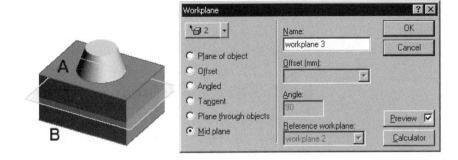

10. Select the Round Frustum feature from the browser pane, right-click, and select Mirror.

11. Set selection mode to workplane and select workplane A indicated in Figure 3–54.

12. Select the OK button.

A mirror copy of the selected feature is constructed. (See Figure 3–55.) Save and close your file (file name: *Mirror.des*).

Figure 3–54
Mirror copy of
a feature being
constructed

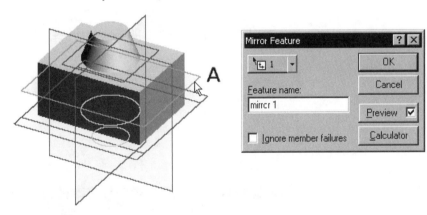

Figure 3–55
Mirror copy of a
feature constructed

Exercises

Now enhance your modeling techniques by completing the following exercises.

Ballpoint Pen Project

To complete the solid parts of the components of the ballpoint pen, you will now add placed solid features to the button, the barrel, and the hand grip.

Button

Now place fillet features on the button of the ballpoint pen.

1. Open the file *BallPen02.des* that you constructed in Chapter 1.

2. Select edges A, B, C, and D indicated in Figure 3–56.

3. Select Feature > Round Edges or select Round Edges from the Design toolbar.

4. In the Round Edges dialog box, set the constant radius to 1 mm and select the OK button.

5. The solid model is complete. Save and close your file.

Figure 3–56
Edges being
rounded

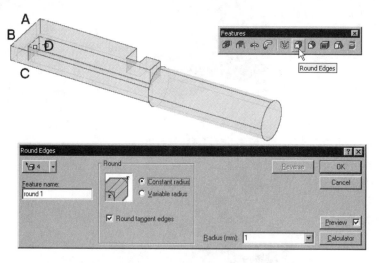

Figure 3–57
Edges rounded

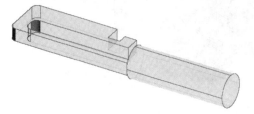

Hand Grip

Construct round edges on the hand grip of the ballpoint pen.

1. Open the file *BallPen05.des* that you constructed in Chapter 1.

2. Construct round edges of 2 mm radius on edges indicated in Figure 3–58.

3. Save and close your file.

Figure 3–58
Round edges being
constructed

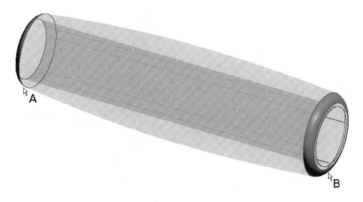

Barrel

Construct round edges and chamfer edges features and add a mirror feature on the solid part of the barrel of the ballpoint pen.

1. Open the file *BallPen03.des* that you constructed in Chapter 1.

2. With reference to Figure 3–59, construct round edges. Radius A = 1 mm, Radius B = 1.5 mm, Radius C = 0.5 mm, and radius D = 0.7 mm.

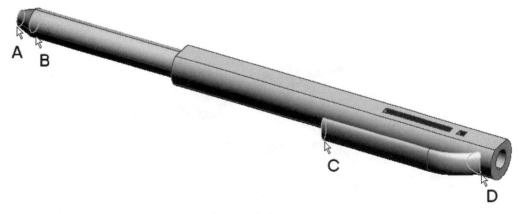

Figure 3–59 Edges to be rounded

3. Select edge A indicated in Figure 3–60 and select Feature >
Chamfer or select Chamfer from the Design toolbar.

4. In the Chamfer dialog box, select equal setback, set
setback to 1 mm, and select the OK button.

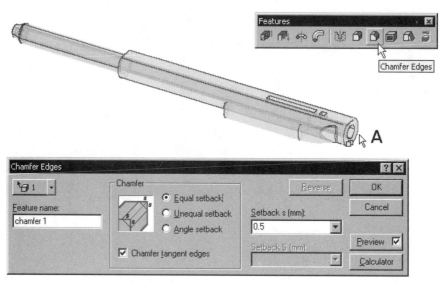

Figure 3–60 Chamfer being placed

5. Select the projection feature highlighted in Figure 3–61
(projection 1), right-click, and select Mirror.

6. In the Mirror dialog box, set selection mode to workplane.

7. Select the base workplane and select the OK button from
the Mirror dialog box.

8. The solid model is complete. Save and close your file.

*Figure 3–61
Mirror feature being
constructed*

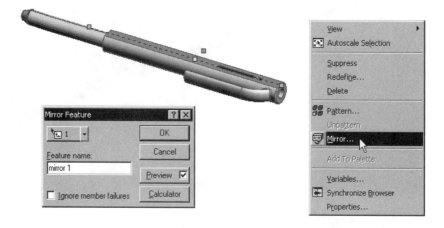

Toy Plane Project

Now you will continue to work on the components of the toy plane.

Upper Body Structure

Follow the steps below to complete the upper body structure of the toy plane by inserting two holes.

1. Open the file *Plane01.des* that you constructed in Chapter 2.

2. With reference to Figure 3–62, construct a sketch on a face of the solid part.

3. In the sketch, construct two circles and add dimensions to set the location of the circles. Note that the size of the circle is unimportant.

*Figure 3–62
Sketch constructed
on the top face of
the solid part*

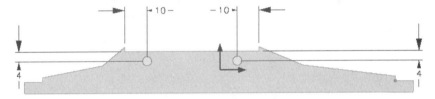

4. Select Feature > Insert Holes or select Insert Holes from the Design toolbar.

5. Select the circles. In the Insert Hole dialog box, select Thru entire part and Simple hole, set the diameter to 4 mm, and select the OK button.

6. The solid part is complete. Save and close your file.

*Figure 3–63
Hole being
constructed*

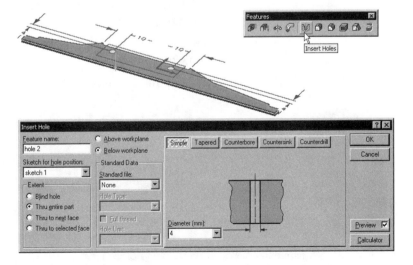

Wing

Complete the wing of the toy plane by inserting two holes.

1. Open the file *Plane12.des* that you constructed in Chapter 2.

2. Construct a sketch on a face of the solid part in accordance with Figure 3–64.

3. Select Feature > Insert Holes or select Insert Holes from the Design toolbar.

4. In the Insert Hole dialog box, select Thru entire part and Simple hole, set the diameter to 4 mm, and select the OK button.

5. Save and close your file.

Figure 3–64
Sketch constructed
on a face of the
solid part

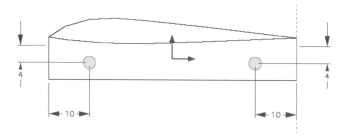

Wheel

Complete the wheel by adding round edge feature.

1. Open the file *Plane08.des* that you constructed in Chapter 2.

2. With reference to Figure 3–65, select the highlighted edges and select Feature > Round Edges or select Round Edges from the Design toolbar.

3. In the Round Edge dialog box, set the radius to 1 mm and select the OK button.

4. The model is complete. Save and close your file.

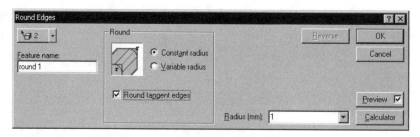

Figure 3–65 Round edges being placed

Wheel Cap

Complete the wheel by adding a round edge feature.

1. Open the file *Plane09.des* that you constructed in Chapter 2.

2. Select the edge highlighted in Figure 3–66 and select Feature > Round Edges or select Round Edges from the Design toolbar.

3. In the Round Edge dialog box, set the radius to 2 mm and select the OK button.

4. The model is complete. Save and close your file.

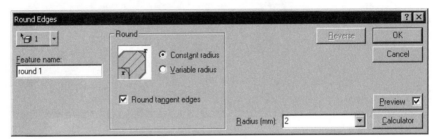

Figure 3–66 Round edge being constructed

Propeller

Complete the propeller of the toy plane by adding a pattern.

1. Open the file *Plane06.des* that you constructed in Chapter 2.

2. Select feature A indicated in Figure 3–67, right-click, and select Pattern.

3. In the Pattern dialog box, set selection mode to edges and select edge B indicated in Figure 3–67.

4. Set the number of instances to 2 and the angle between adjacent instances to 180°.

5. Select the OK button.

6. The propeller is complete. Save and close your file.

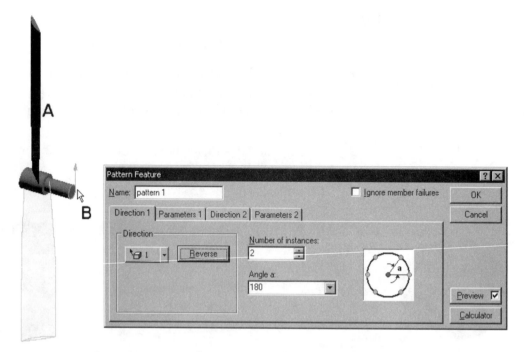

Figure 3–67 Pattern being constructed

Rear Mounting

Add a round edge to the rear mounting of the toy plane.

1. Open the file *Plane05.des* that you constructed in Chapter 2.

2. Select the edge highlighted in Figure 3–68 and select
   Feature > Round Edges or select Round Edges from the
   Design toolbar.

3. In the Round Edge dialog box, set the radius to 3 mm and
   select the OK button.

4. The model is complete. Save and close your file.

Figure 3–68 Round edge being placed

Front Mounting

Complete the front mounting of the toy plane by adding a shell feature and a couple of sketched features.

1. Open the file *Plane03.des* that you constructed in Chapter 2.

2. Select faces A and B indicated in Figure 3–69 and select Feature > Shell Solids or select Shell Solids from the Design toolbar.

3. In the Shell Solids dialog box, select Outside, set offset to 1 mm, and select the OK button.

*Figure 3–69
Shell feature being
constructed*

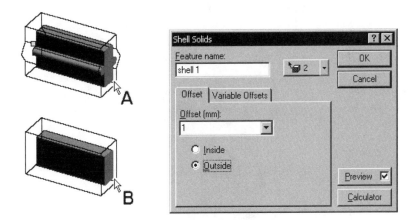

4. Construct a sketch on the frontal workplane as shown in Figure 3–70.

5. Extrude the sketch 1 mm symmetric about the workplane to add material to the solid part. (See Figure 3–71.)

*Figure 3–70
Sketch constructed
on the frontal
workplane*

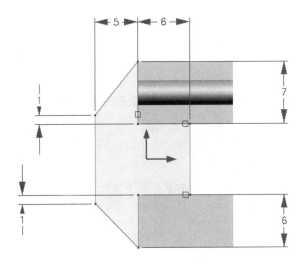

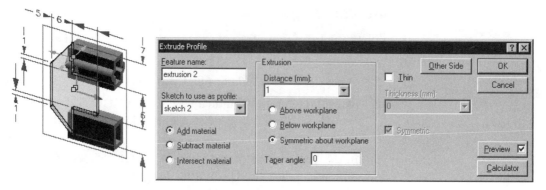

Figure 3–71 Sketch being extruded

6. Construct a sketch on a face of the solid part in accordance with Figure 3–72.

Figure 3–72
Sketch constructed

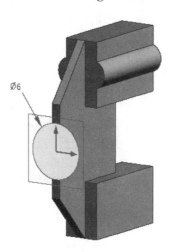

7. Project the sketch below the workplane to face A indicated in Figure 3–73 to add material to the solid part.

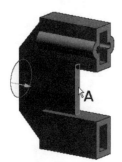
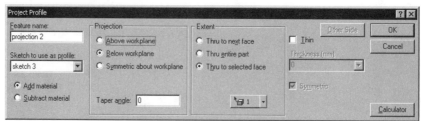

Figure 3–73 Sketch being projected

8. Construct a workplane on face A indicated in Figure 3–74 and construct a sketch on the workplane.

9. Construct a circle to specify a center position for inserting a hole.

10. Insert a simple hole of 4 mm to cut through the solid part.

11. The solid part is complete. Save and close your file.

Figure 3–74
Center of hole
constructed

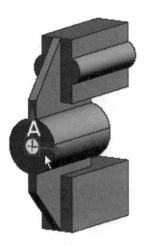

Summary

Constructing feature-based solid models concerns making unique, modular solid features one by one and combining them as you construct them. Besides composing a solid model from four basic kinds of sketched solid features (extruded, revolved, swept, and lofted), which you learned in the previous chapter, you can incorporate preconstructed solid features in your design such as rounded edge, chamfered edge, face draft, shell body, cosmetic threads, and insert hole. Except for insert hole, which requires a sketch to depict the hole centers, the other kinds of pre-constructed solid features do not require sketching. You only need to select them from the menu and specify appropriate parameters.

To help reduce design lead time, you can drop sketches, features, and components into your design from the palette, which is a customizable standard library. To reuse sketches, features, and components already constructed, you can put them in the palette for future use. In addition to adding new library objects to the palette, you can delete unwanted objects from the palette.

Features in a solid model can be repeated by constructing a pattern or a mirror copy. A pattern duplicates selected features in one or two directions. If the features are repeated in one linear direction, a row of features is formed. If the features are repeated in two linear directions (not necessarily perpendicular to each other), a pattern in the form of a parallelogram is formed. If the direction of repetition is a circular path, a circular pattern is formed. If one direction is linear and the other direction is circular, a circular pattern with multiple rings is formed. A mirror produces a mirror copy of selected features. By default, the instances of a pattern or the mirror copy are linked to the source features. If you re-define the source, the pattern and the mirror copy will change as well. In making a pattern, you can apply an incremental size change to the feature and the sketches of the source features, and you can also unpattern a pattern so that you can change the parameters of the duplicated features individually.

Grouping is a means of treating a set of features collectively. When you make a pattern or a mirror copy, selected source features are first put into a group before the pattern or mirror operation.

Review Questions

1. State, with the aid of sketches, the kinds of pre-constructed solid features that you can incorporate into a solid model.

2. Explain how to make use of the features and sketches available from the palette in solid modeling.

3. How are new objects incorporated into the palette and unwanted objects removed from the palette?

4. In making a pattern, how many ways can you duplicate a feature or a set of features?

5. What kind of objects will you use as a mirror plane for making a mirror copy of a feature or a set of features?

6. If an individual instance of a pattern or a mirror copy fails to construct properly, pattern and mirroring operation will cease. What will you do to overcome this problem?

7. How are the grouping and ungrouping of features carried out?

CHAPTER 4

Assembly Modeling

Objectives

This chapter delineates the concept of assembly modeling, detailing approaches in making the assembly of a product or system in the computer. Product structure and color and material setting are also explained. After studying this chapter, you should be able to

❏ Explain the key concepts of assembly modeling

❏ Use three design approaches in constructing the assembly of a product or system

❏ Set the color and materials of the components of an assembly

Overview

An assembly of components is a collection of components put together to form a product or a system. The task of making an assembly model in the computer involves two major tasks: gathering together a set of components and maintaining proper positional relationships among the components. There are three approaches to designing a product or a system. In the first approach, you start from making the individual components and work upward to build the assembly. In the second approach, you start from the assembly as a whole and work downward to construct individual components. In the third approach, you construct some components and work upward as well as work downward to construct individual components in the context of the assembly.

Assembly Modeling Concepts

With the exception of very simple objects, such as a ruler, most products and systems have more than one part put together to form a useful whole. The set of parts put together is called an assembly. When you design the parts of an assembly, the relative dimensions and positions of parts, and the way they fit together, are crucial. You also need to know whether there is any interference among the mating parts. If there is interference, you need to find out where it occurs, and then you can eliminate it. To shorten the design lead time, you construct virtual assemblies in the computer to validate the integrity of the parts and the assembly as a whole.

Definition

For complex products and systems that have many parts, it is common practice to organize the parts into a number of smaller subassemblies such that each subassembly has fewer parts. Therefore, an assembly set depicting a product or system may consist of a number of parts or a number of parts together with a number of subassemblies. Collectively calling the individual parts and the subassemblies as components, you can define an assembly in the computer as a data set consisting of information about a collection of components linked to the assembly and about the way the components fit together in the assembly. Figure 4–1 shows a set of components assembled together to form a virtual assembly in the computer.

Figure 4–1
Virtual assembly of
components in the
computer

Assembly Hierarchy

The hierarchy of an assembly has an inverted tree structure, consisting of the assembly of the product or system at the top and the subassemblies and parts below. Figure 4–2 shows the assembly hierarchy tree of the toy plane shown in Figure 4–1.

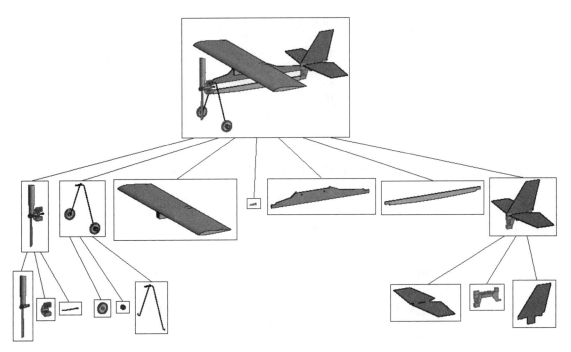

Figure 4–2 Assembly hierarchy tree

As shown in Figure 4-2, the top of the hierarchy tree is the complete assembly of the toy plane. Down below in the hierarchy tree are, from left to right, the propeller subassembly, the landing gear subassembly, the wing, the wing locking pin, the upper body structure, the lower body structure, and the tail subassembly. Note that even though there are two wing locking pins in the final assembly, we will consider it as one component in the hierarchy. Any duplication of components in an assembly is called an instance.

Further down in the hierarchy tree are the components of three subassemblies. The propeller subassembly has three components: the propeller, the front mounting, and the drive shaft. The landing gear subassembly also has three components: the landing gear main structure, the wheel, and the wheel cap. Again, two wheels and two wheel caps are used in the landing gear subassembly, and they are treated as instances. The final subassembly is the tail subassembly. It has three components: the horizontal stabilizer, the rear mounting, and the vertical stabilizer.

Assembly Modeling Tasks

Constructing a virtual assembly of a product or system in the computer consists of two major tasks:

1. Linking together a set of files, each depicting the solid model of individual components

2. Establishing positional relationships among the components

Linking Component Design Files to an Assembly Design File

Using Pro/DESKTOP, you use a design file to make an assembly of a product or system as well as to construct individual solid parts, as you did in Chapter 2 and 3.

Therefore, your first task in making an assembly is to start a design file, use it as an assembly file, and link to it a set of design files, each depicting a component of the assembly, which can be a single part or another assembly. If the link component is an assembly, the design file will, in turn, have to link to a set of design files. Figure 4–3 shows the relationships among various design files used for making the assembly of the toy plane shown in Figure 4–1. Ways to link component design files to an assembly design file together with how to manipulate components in an assembly will be discussed later in this chapter.

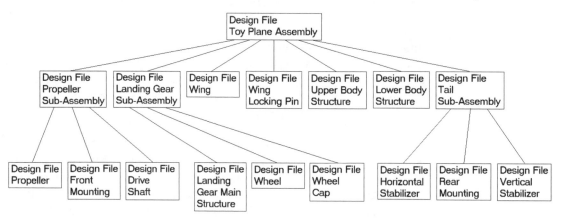

Figure 4–3 Relationships among the design files used for different purposes

Establishing Position Relationships Among Components in an Assembly

After linking individual components' design files to the assembly design file, you will see a number of components in the work pane of the assembly design file. Initially, each component is free to translate in such a way analogous to an object floating in the outer space. You can select it and drag it to any location. To restrict the free movement and establish positional relationships among the components in an assembly, you perform the second task, to set up mating conditions. Figure 4–4 shows two components in an assembly design file, and Figure 4–5 shows two mating conditions applied, causing a pair of faces to mate and a pair of axes to orient. Details regarding how to apply mating conditions will be discussed later in this chapter.

Figure 4–4
Two components free to translate

Figure 4–5
Relative movement between components restricted by mating conditions

Update and Propagate

Because the individual components' design files are linked (not copied) to the assembly design file, individual components' data is loaded into the computer's memory each time you open the assembly design file. As a result, the most up-to-date information about the components is always loaded.

If the assembly design file is already opened and the individual component's design file is modified, you can force an immediate update to the assembly by selecting Feature > Update and Propagate.

Design Approaches

Ways to link individual component's design file to an assembly's design file depend on which approach you use to design and construct an assembly of a product or system: the bottom-up approach, top-down approach, or hybrid approach.

Bottom-Up Approach

As the name implies, the bottom-up approach starts from making the bottom objects of the assembly hierarchy tree, which are the individual solid parts. After making the parts at the bottom of the hierarchy tree, you move upward to construct the subassemblies and then the final assembly.

When you already have a good idea on the size and shape of the components of a product or system or you are working as a team on an assembly, you use the bottom-up approach. Through parametric solid modeling methods, you construct all the parts to appropriate sizes and shapes that best describe the components of the assembly. Then you start a design file, use it as an assembly design file, and add the components you already constructed in the assembly. In the assembly, you align all the components together by applying mating conditions. After putting all the components together, you analyze and improvise your design by making necessary changes to the parts.

Figure 4–6
Selecting Add Component from the Assembly menu

Using the bottom-up approach to add a component design file to an assembly design file, select Assembly > Add Component. (See Figure 4–6.) In the Component Part dialog box, select a design file and select the OK button.

Top-Down Approach

Top refers to the assembly of the product or system. In a top-down approach, you start from an assembly and proceed downward to construct the individual components.

Sometimes you have a concept in your mind, but you do not have any concrete ideas about the components. You use the top-down approach by starting a design file for assembly purposes and constructing a component in the context of the assembly. From the preliminary component parts, you improvise. The main advantages in using this approach are that you see other components while working on an individual component, and you can continuously switch from designing one part to another. The disadvantage in using this approach is that you may be confused about which design file you are working on, the design file of the assembly or the design file of the individual solid part.

Using the top-down approach to construct a part in the context of an assembly, select Assembly > New Design in Context. (See Figure 4–7.)

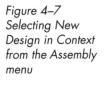

*Figure 4–7
Selecting New
Design in Context
from the Assembly
menu*

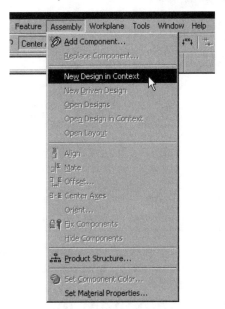

Hybrid Approach

The hybrid approach is a combination of the bottom-up and top-down approaches. In reality, you seldom use one approach alone. You use the bottom-up approach for standard components and new components that you already know about, and you use the top-down approach to figure out new components with reference to the other components.

Manipulating Components in an Assembly

Ways to manipulate components in an assembly include opening a component's file in the context of an assembly, deleting components, hiding components, translating components, duplicating components, setting a component's color, and setting a component's material.

Opening a Component's File in the Context of an Assembly

No matter which approach you use to construct an assembly, you can always open a component of the assembly in the context of the assembly by selecting the component from the component browser pane, right-clicking, and selecting Open in Context. (See Figure 4–8.)

Figure 4–8
Opening a design file of a component in the context of an assembly

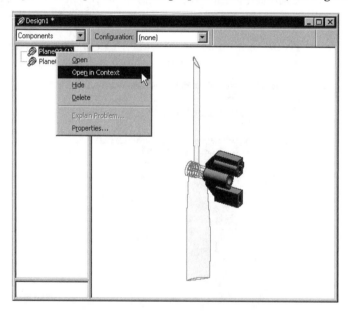

Opening a file in the context of an assembly lets you see other components of the assembly while working on an individual component.

Deleting a Component in an Assembly

If a component is no longer required in an assembly, you delete it from the assembly. It must be noted that deleting a component only removes the link between the component and the assembly. The component is not deleted physically from your computer's hard disk. To delete a component in an assembly, set the browser pane to display component, right-click the component from the browser, and select Delete from the right-click menu shown in Figure 4–8.

Hiding a Component in an Assembly

In a very complex assembly, you may want to make a certain component invisible so that you can concentrate on working on some other components. To hide a component, select Hide from the right-click menu shown in Figure 4–8.

Translating Components

Prior to applying mating conditions to a pair of components, you may want to translate them to an appropriate position in 3D space. Translation of components can be done in three ways. The first way is to select and drag a component to a new location. The second way is to select a component, select Edit > Shift Point-to-Point or right-click and select Shift Point-to-Point, and then select two locations on the screen to depict a vector delineating the direction and distance of translation. In the third way, to translate a

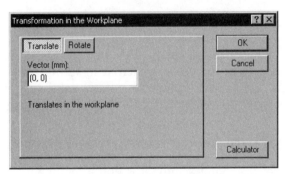

component to a precise location, select the component and select Edit > Transform or right-click and select Transform, and specify the translation vector or the rotation angle in the Transformation in the Workplane dialog box. (See Figure 4–9.)

Figure 4–9 Translation in the Workplane dialog box

Duplicating Components

To duplicate components already placed in an assembly in a

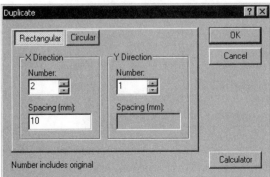

rectangular or circular pattern, select the component, select Edit > Duplicate or right-click and select Duplicate, and specify the parameters of the pattern in the Duplicate dialog box. (See Figure 4–10.)

Figure 4–10 Duplicate dialog box

Setting a Component's Color

To set the color of a component in an assembly, select the component and then select Assembly > Set Component Color. In the Color dialog box shown in Figure 4–11, select a color and select the OK button.

*Figure 4–11
Color dialog box*

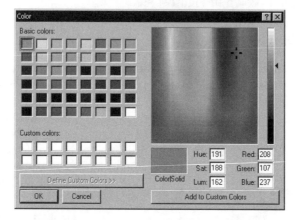

Color specification can be done in three ways. The first way is to select a color from the color swatch. The second way is to specify the hue, saturation (sat), and luminous (lum) values of a color. The third way is to specify the red, green, and blue values of a color.

Setting a Component's Material

To set the material properties of a component in an assembly, select the component and select Assembly > Set Material Properties. In the Material Properties dialog box shown in Figure 4–12, select a material and select the OK button.

*Figure 4–12
Material Properties
dialog box*

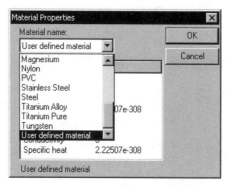

Material properties can be specified in two ways. The first way is to select a material from the Material name dropdown list of the Material Properties dialog box. The second way is to select User defined material in the Material name list and specify material properties.

Applying Mating Conditions

To restrict the movement of a component in 3D and to align it with another component in the assembly, you apply mating conditions to selected edges and faces. There are six ways to apply mating conditions: align, mate, offset, center axes, orient, and fix component.

Align Condition

The align condition causes selected planar faces or workplanes to align with each other on a common planar face with the normal directions of the selected faces pointing in the same direction. It also sets up a tangent relationship between selected extruded faces (including extruded ellipses and extruded splines) and swept faces. To apply an align condition, select a pair of faces and then select Assembly > Align. Figure 4–13 shows two planar faces A and B aligned.

Figure 4–13 Align condition applied to two planar faces

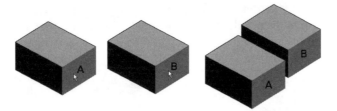

If you want to maintain a separation distance between the selected faces, use offset condition instead of align condition. To set up an angular separation, use orient condition instead.

Mate Condition

The mate condition causes selected planar faces and workplanes to mate together on a common planar face with the normal directions of the selected faces pointing toward each other. It also sets up a tangent relationship between selected extruded faces (including extruded ellipses and extruded splines) and swept faces. To apply a mate condition, select a pair of faces and then select Assembly > Mate. Figure 4–14 shows two planar faces A and B mated.

Figure 4–14 Mate condition applied to two planar faces

To maintain a separation distance between mated faces, use offset condition. To set up an angular separation, apply orient condition.

Offset Condition

The offset condition has two options: align and mate, maintaining a separation between aligned or mated faces. Apart from an added separation distance, an offset condition is the same as an align condition or a mate condition. To apply an offset condition, select a pair of faces and then select Assembly > Offset. Figure 4–15 shows an offset condition applied to two planar faces A and B.

*Figure 4–15
Offset mate
condition applied to
two planar faces*

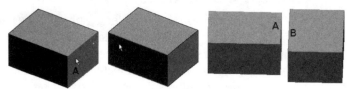

Center Axes Condition

The center axes condition sets up a concentric relationship between the center points or axes of selected circular edges, selected circular faces, selected axial features, or selected conical faces. It also causes selected linear edges to be collinear. To apply a center axes condition, select a pair of circular edges, circular faces, axial features, conical faces, or straight edges and then select Assembly > Center Axes. Figure 4–16 shows a center axes condition applied to a pair of circular edges A and B, causing the axes to align, and Figure 4–17 shows a center axes condition applied to a pair of straight edges C and D, causing the edges to align.

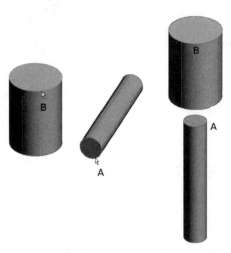

Figure 4–16 Center axes condition applied to two circular edges

*Figure 4–17
Center axes
condition applied to
two straight edges*

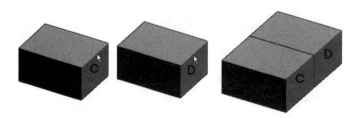

Orient Condition

The orient condition causes selected faces, linear edges, axes, or workplanes to maintain at a separation angle. To apply an orient condition, select a pair of faces, linear edges, axes, or workplanes and then select Assembly > Orient. Figure 4–18 shows an orient condition applied to a pair of planar faces A and B and Figure 4–19 shows an orient condition applied to a pair of edges C and D.

*Figure 4–18
Orient condition
applied to a pair of
planar faces*

*Figure 4–19
Orient condition
applied to a pair of
edges*

Fix Components Condition

The fix components condition causes selected components to be fixed in 3D space—they cannot be moved. To apply a fix components condition, select one or more components and select Assembly > Fix Components.

Manipulating Mating Conditions

Mating conditions already applied in an assembly can be manipulated in several ways: delete, suppress, and modify.

Deleting a Mating Condition

To delete unwanted mating conditions, set the browser pane to display components in the design file, expand the Contraints folder at the bottom of the browser pane, right-click a mating condition you wanted to delete, and select Delete. (See Figure 4–20.)

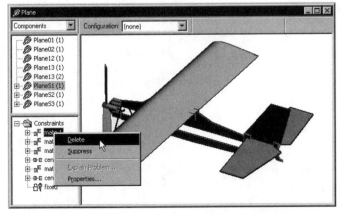

Figure 4–20 Deleting mating condition

Suppressing a Mating Condition

A mating condition deleted from the browser pane is permanently removed. If you want to see what will happen if a mating condition is removed with the option of reinstating the condition, you suppress it instead of deleting it. In the right-click menu shown in Figure 4–20, select Suppress.

Modifying a Mating Condition

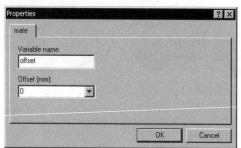

To modify the parameters of a mating condition, select Properties from the right-click menu shown in Figure 4–20. In the Properties dialog box, change the parameters as may be appropriate. (See Figure 4–21.)

Figure 4–21 Properties dialog box of a mating condition

Check Interference

In reality, a solid object cannot be put inside another solid object without breaking or removing part of the materials. However, in the computer, you can place a component wholly or partly inside another component without any warning, unless you perform an interference check.

To check interference among the components in an assembly, select the components and select Tools > New Measurement > Interference. Information regarding interference among the components will be listed in the Interference List dialog box. (See Figure 4–22.) Note:

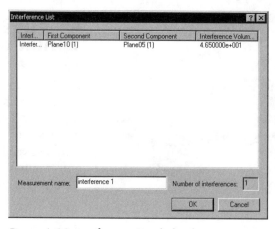

If you click the OK button in the Interference List dialog box, a new measurement is created in the system that constantly updates the value of the interference volume. (Measurements are discussed in a later chapter.) If your intent is simply to check if there is any interference, then click the Cancel button to dismiss the dialog box.

Figure 4–22 Interference List dialog box

Product Structure

A product structure of an assembly depicts the assembly hierarchy tree in a tabulated format. To generate a model tree report for an assembly of a product or system, you construct a product structure report by selecting Assembly > Product Structure.

In the Product Structure dialog box, you can obtain a parts list, delineating only the first level components in making the assembly, as well as a bill of materials, listing all the components used in making the assembly. (See Figures 4–23 and 4–24.)

Figure 4–23
Product Structure
dialog box showing
parts list

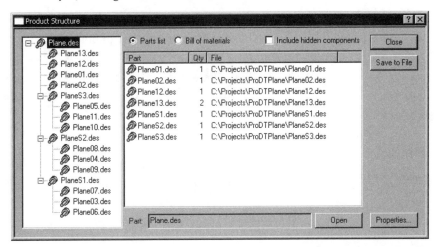

Figure 4–24
Product Structure
dialog box showing
bill of materials

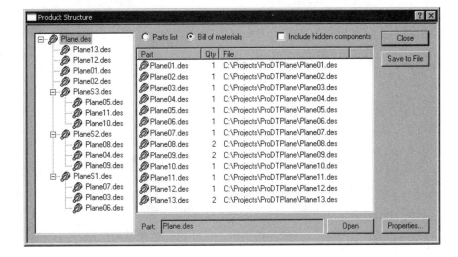

Bottom-Up Approach—Ballpoint Pen Project

Because all the components of the ballpoint pen are already complete, you will use the bottom-up approach to put them together in an assembly.

1. Start a new design file and set the measurement units to millimeters.

Now you will link up two components' design files to the current assembly design file and apply two assembly mating conditions to set their positional relationship.

2. Select Assembly > Add Component.

3. In the Component Part dialog box, select the file *BallPen01.des*. The file is linked to the current file.

4. Repeat steps 2 and 3 to link the file *BallPen02.des*. Two files are linked. (Note: A shortcut to linking the files is to drag the files from the Windows Explorer window into the work pane of the design window.)

5. Select edges A and B indicated in Figure 4–25 and select Assembly > Center Axes. (Note: Press the SHIFT key while selecting multiple objects.)

The axes of the components are aligned.

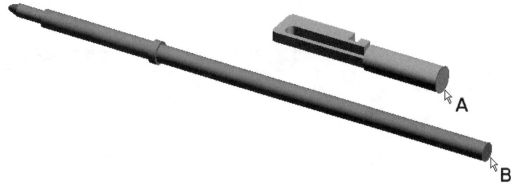

Figure 4–25 Two files linked and axes being aligned

6. Select face A of Figure 4–26, rotate the display, and select face A of Figure 4–27.

7. Select Assembly > Mate.

The faces are mated.

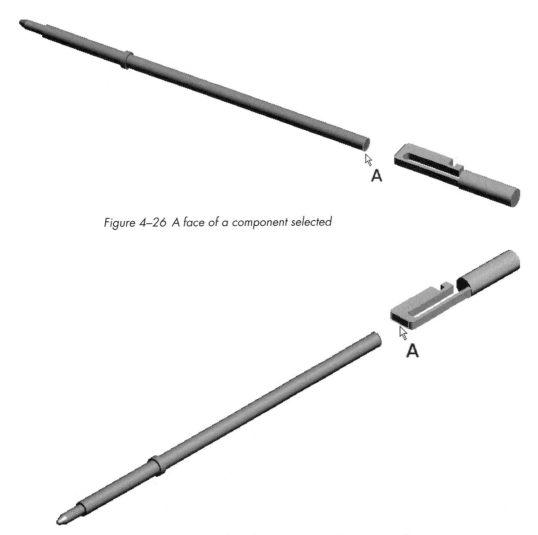

Figure 4–26 A face of a component selected

Figure 4–27 A face of the other component selected as well

Now link the spring's design file (*BallPen04.des*) and apply assembly constraints to the position the spring. As we have mentioned in Chapter 2, two revolved solid features, one added to the solid and the other one intersected with the solid, are incorporated to facilitate assembly. Here you will apply constraint to a circular edge and two flat faces of the spring.

8. Set the display to transparent, select flat face A of the pen tip and flat face B of the spring indicated in Figure 4–28, and select Assembly > Mate.

The selected faces are mated.

9. Zoom in the display, select circular edges A and B indicated in Figure 4–29, and select Assembly > Center Axes.

The axis of the spring is aligned to the axis of the pen tip.

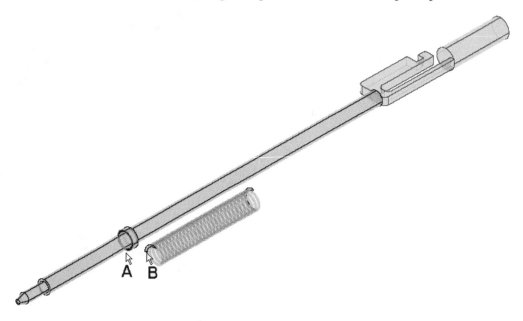

Figure 4–28 Faces being mated

Figure 4–29
Circular edges
selected

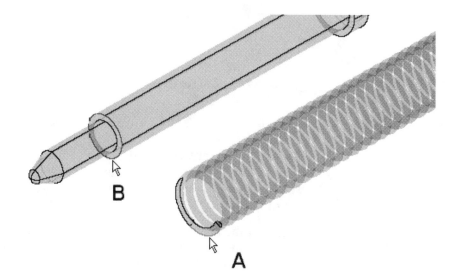

Now link the barrel's design file to the assembly design file and add assembly constraints to position it.

10. Select Assembly > Add Component and select the file *BallPen03.des*.

11. Select the flat face A of the spring indicated in Figure 4–30, rotate and zoom the display, select flat face A indicated in Figure 4–31, and select Assembly > Mate.

The selected faces are mated together.

Figure 4–30
Flat face of the
spring selected

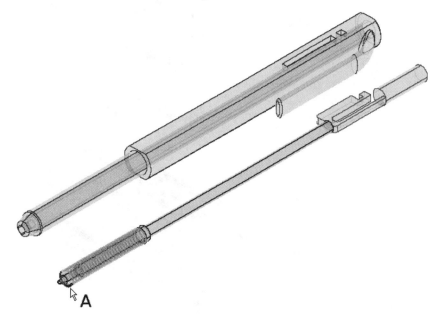

Figure 4–31
A flat face inside
the ball pen barrel
selected

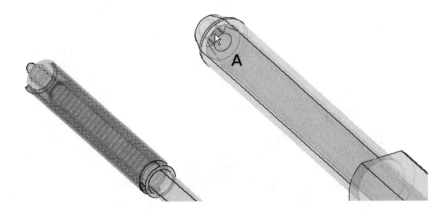

12. Set the display to shaded, rotate and zoom the display, select faces A and B indicated in Figure 4–32, and select Assembly > Mate.

Figure 4–32
Faces being mated

13. Select circular edges A and B indicated in Figure 4–33 and select Assembly > Center Axes.

Figure 4–33
Circular edges
selected

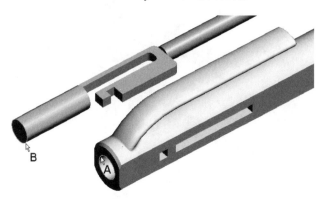

Now put the hand grip into the assembly.

14. Put the component BallPen05 into the assembly design file.

15. Select faces A and B indicated in Figure 4–34 and select Assembly > Mate.

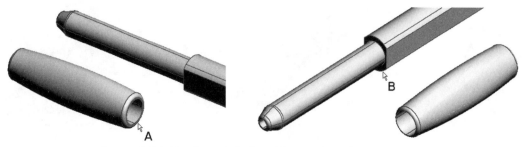

Figure 4–34 Hand grip linked and faces selected

16. Select circular edges A and B indicated in Figure 4–35 and select Assembly > Center Axes.

Figure 4–35
Circular edges
selected

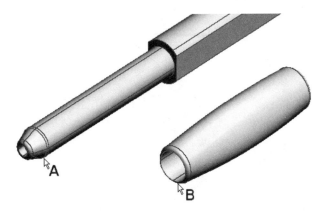

17. The assembly is complete. (See Figure 4–36.) Save and close your file (file name: *BallPen.des*).

Figure 4–36
Completed assembly

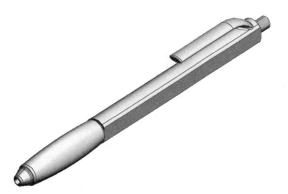

Hybrid Approach—Toy Plane Project

In reality, we usually construct some components we already know about by using the bottom-up approach and construct matching components by using the top-down approach. Now you will construct the toy plane assembly by using the hybrid approach.

1. Start a new design file and set the measurement units to millimeters.

2. Select Assembly > Add Component and select the file *Plane12.des* to link it to the design file.

3. Select the component, right-click, and select Fix Component.

The component is fixed in 3D space.

4. Link the file Plane01.des into the design file. (See Figure 4–37.)

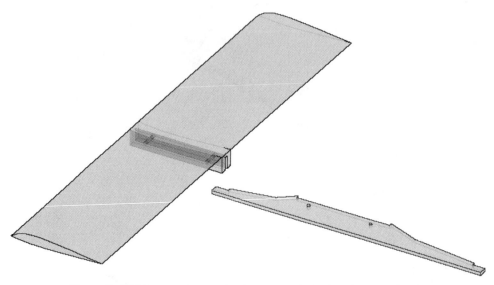

Figure 4–37 Wing and upper body structure linked to the assembly

5. Rotate and zoom the display, select faces A and B indicated in Figure 4–38, and select Assembly > Mate.

The selected faces are mated.

*Figure 4–38
Faces being mated*

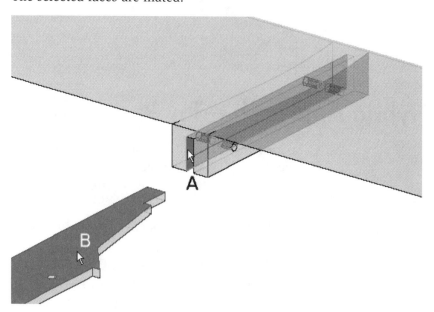

6. Rotate and zoom the display, select circular edges A and B indicated in Figure 4–39, and select Assembly > Center Axes.

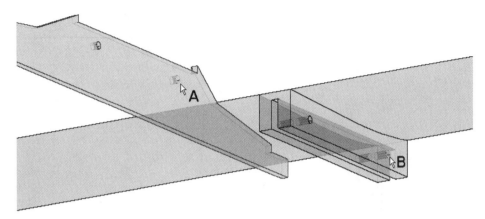

Figure 4–39 Center axes being aligned

7. Select and drag the edge of the wing to rotate it about the aligned axes.

8. Select edges A and B indicated in Figure 4–40 and select Assembly > Orient.

9. In the Orient dialog box, set the angle to 180° and select the OK button.

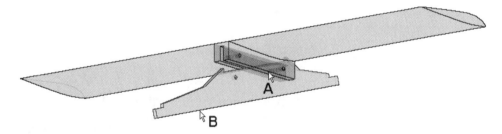

Figure 4–40 Component dragged and center axes being aligned

The components are properly assembled.

10. Save your file (file name: *Plane.des*).

Now you will construct the wing locking pin in the context of the assembly.

1. Select Assembly > New Design in Context.

2. With reference to Figures 4–41 and 4–42, construct a sketch and revolve the sketch about line A.

Figure 4–41
New design in
context

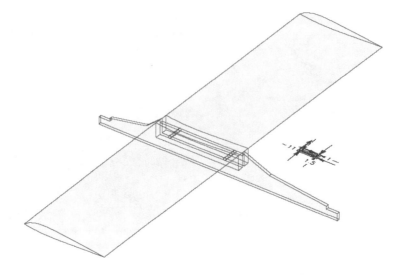

Figure 4–42
Sketch being
revolved

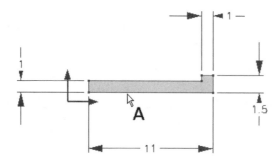

3. Save and close the file in the context of the assembly (file name: *Plane13.des*).

Now you are back to the assembly design file.

4. Select faces A and B indicated in Figure 4–43 and select Assembly > Mate.

Figure 4–43
Flat faces being
mated

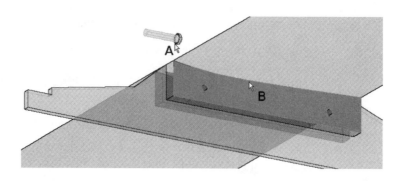

5. Select circular edges A and B indicated in Figure 4–44 and select Assembly > Center Axes.

Figure 4–44
Center axes being
aligned

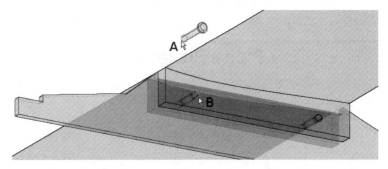

6. With reference to Figure 4–45, add a second wing locking pin and apply assembly constraints to properly position it. (Note: You can duplicate the wing locking pin by selecting Edit > Duplicate.)

Figure 4–45
Second locking
pin placed and
assembled

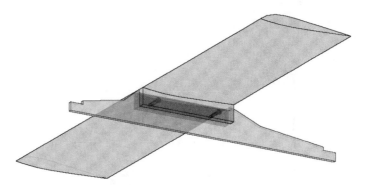

Two components are linked to the assembly and a component is constructed in the context of the assembly. Now save and close your file.

Modifying Design in Context

Now matter which design approach you use to construct an assembly, you can always modify any component in the context of the assembly.

To modify a component in the context of an assembly, set the browser pane to display components, right-click the component from the browser, and select Open in Context. The selected component's design file will be opened in the context of an assembly. In the work pane, you will see other components of the assembly while working on the component.

If you select Open in the right-click menu instead, you will open the component in a usual way.

Exercises

Now perform the following exercises to enhance your assembly modeling knowledge.

Toy Plane Project

Now complete the toy plane project by first constructing two more subassemblies and then construct a final assembly.

Propeller Subassembly

With reference to Figure 4–46, construct an assembly of the propeller, front mounting, and drive shaft. After construction, save and close the file (file name: *PlaneS1.des*).

*Figure 4–46
Propeller
subassembly*

Landing Gear Subassembly

Construct an assembly with a landing gear main structure, two wheels, and two wheel caps in accordance with Figure 4–47 (file name: *PlaneS2.des*).

*Figure 4–47
Landing gear
subassembly*

Tail Subassembly

Referring to Figure 4–48, construct an assembly consisting of the rear mounting, horizontal stabilizer, and vertical stabilizer (file name: *PlaneS3.des*).

Figure 4–48
Tail subassembly

Final Assembly

Open the Plane assembly file, put the propeller subassembly, the landing gear subassembly, the tail subassembly, and the lower body structure into the assembly, and add assembly constraints to properly assemble the components in accordance with Figure 4–49.

Figure 4–49
Plane assembly

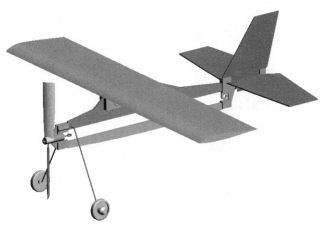

Summary

In Pro/DESKTOP, a design file can be used for two purposes: constructing an individual component or constructing an assembly model. To depict a product or system in the computer, you construct an assembly model in a design file.

Assembly modeling consists of two major tasks: linking a set of components saved in a set of individual design files to the assembly design file and applying mating conditions to selected edges, faces, and features of the components in the assembly design file to establish positional relationships.

There are three design approaches to setting up an assembly: the bottom-up approach, top-down approach, and hybrid approach. Because an assembly hierarchy is an inverted tree with the assembly of the product or system at the top and the components at the bottom, bottom refers to the individual components and top refers to the final assembly. In a bottom-up approach, you construct individual components and then construct an assembly model by linking the components to the assembly. In the top-down approach, you construct individual components in the context of an assembly so that you see other components while designing. In reality, you usually use a hybrid approach, which is a combination of the bottom-up approach and the top-down approach.

After linking components together to form an assembly, you apply mating conditions to restrict components' movement and positional relationships, using align, mate, offset, center axes, orient, and fix components conditions.

In an assembly, a component can be removed by deleting and can be hidden from the work pane so that it is not visible. Mating conditions can be deleted, suppressed, and modified. You can also set the color and material properties of components and generate a product structure.

Review Questions

1. What are the purposes of making an assembly model in the computer?

2. What are the two major tasks in assembly modeling?

3. Delineate the three design approaches in making an assembly of a product or system.

4. State the kinds of mating conditions that you will apply to the components of an assembly to restrict the components' movement and set components' positional relationship.

CHAPTER 5

Advanced Modeling Techniques

Objectives

This chapter explains how to construct a solid part from an assembly of solid parts and delineates ways to modify the faces and body of a solid model and to treat imported objects. Various advanced design tools are also explained. After studying this chapter, you should be able to

- ❒ Construct a solid part in a non-linear way

- ❒ Modify a solid part's face and body

- ❒ Treat imported objects

- ❒ Use various modeling tools

Overview

The normal way of constructing a solid part is to construct its features in a linear way, in which each feature has to be combined with the solid part as it is constructed. To overcome this limitation, you can add one or more individual design files to a solid part and combine the solids into a single solid.

Beyond using the editing tools depicted in Chapter 2, you can modify a solid's face by deformation, removal, transformation, offset, and replacement. (These editing methods are particularly useful for treating imported solids without parametric history.) In addition, you can mirror, scale, and trim a solid part and convert a feature-based parametric solid part to a static part with a single feature. With imported surface or quilt, you can thicken them to produce a solid.

As a concluding chapter on modeling, this chapter also addresses various design tools, including variables, design rules, configurations, animation, solver, dimension slider, and measurement.

Linear and Non-Linear Modeling Approach

The normal way to make a solid part is to construct its solid features one by one and combine them as you construct them. This way of modeling a solid part is called the linear approach. (See Figure 5–1.)

Figure 5–1
Linear approach in
modeling

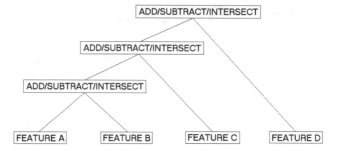

The major disadvantage of this approach is that you cannot first combine solid features into two or more sets of features and then combine the sets of features into a single part. For example, you cannot join features A and B and features C and D and then cut the combined C and D features from the combined A and B features. As a result, the final form and shape of the solid part is restricted.

Non-Linear Modeling Approach

To overcome this drawback, you can decompose a complex solid part into two or more sets of solid features, construct each set of features in separate, individual design files, put the solid parts together in a single file, and combine them into a single solid part. This method of modeling is called the non-linear modeling approach. (See Figure 5–2.)

Figure 5–2
Non-linear modeling
approach

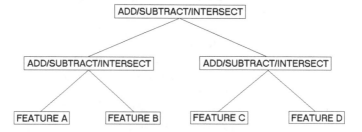

Use Components

To construct a solid part using the non-linear approach, you first construct two or more solid parts in separate design files, depicting two or more sets of solid features of the complex solid part. Then you open one of the design files, add the other design files as if you are working on an assembly, and add appropriate mating conditions to properly locate the solid parts. Finally, you select Features > Use Components to use the component as a single feature and combine it with the solid part.

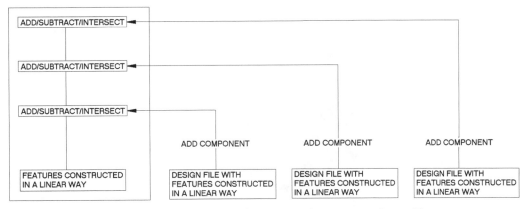

Figure 5–3 Combining solids constructed in separate design files

Now perform the following steps to learn how to construct the features of a solid part in a non-linear way. You will first construct two design files, then add one of the files to the other file, and use the linked file as a tool for making a complex solid.

1. Start a new design file and set the measurement units to millimeters.

2. Double-click the Block from the Base Shapes tab of the palette to drop it onto the base workplane.

3. Modify the block's dimensions to 160 mm times 120 mm.

4. Shell the solid with the top face removed. Shell thickness is 10 mm. (See Figure 5–4.)

5. Save the file (file name: *Nonlinear1.des*).

Figure 5–4
Block's size
modified and
shelled

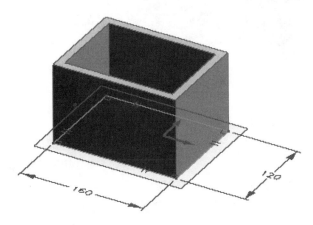

6. Modify the block's dimension to 100 mm times 80 mm. (See Figure 5–5.)

7. Select File > Save Copy As to save the file as *Nonlinear2.des* and close the current file.

Now you have two shelled blocks with different dimensions.

Figure 5–5
Second design file

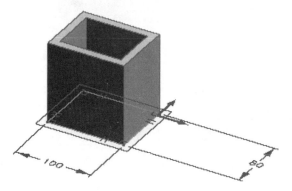

Now construct an assembly of the two components.

1. Start a new design file and set the measurement units to millimeters.

2. Select Assembly > Add Component.

3. Select the file *Nonlinear1.des* and select the OK button.

4. Select Assembly > Add Component and put the file Nonlinear2.des into the current file. (See Figure 5–6.)

Figure 5–6
Two design files
added to the current
design file

Now apply assembly constraints.

1. Select faces A and B indicated in Figure 5–7. (Note: You can move the Nonlinear2.des a little bit so that selection becomes easier.)

2. Select Assembly > Offset.

3. Select Align, set offset distance to 20 mm, and select the OK button.

Figure 5–7
Offset constraint
being applied

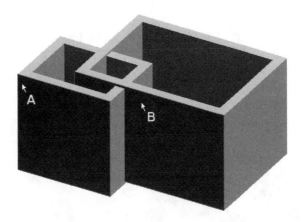

4. Align faces A and B indicated in Figure 5–8 with an offset distance of -100 mm.

5. Align faces A and B indicated in Figure 5–9.

Figure 5–8
Offset constraint
being applied

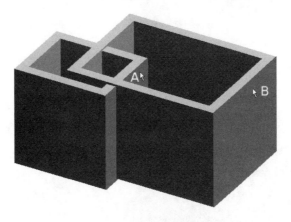

Figure 5–9
Align constrained
applied

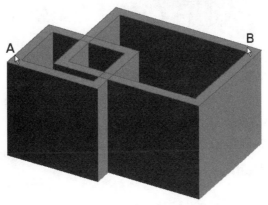

Now use the link components as tools to construct a complex solid.

1. Select the component Nonlinear1 and select Feature > Use Component.

2. In the Use Component dialog box, select the Add material button and select the OK button.

3. Select the component Nonlinear2 and select Feature > Use Component.

4. Select the Add material button and select the OK button.

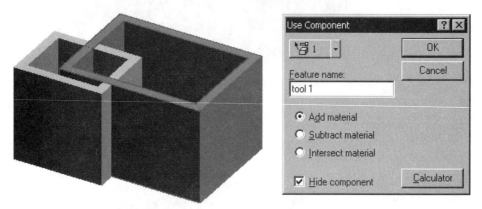

Figure 5–10 Adding a component to the design file

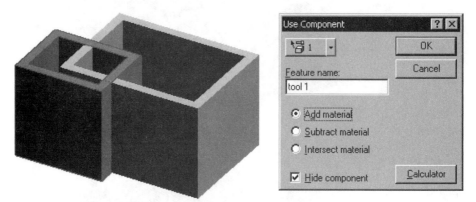

Figure 5–11 Adding a component to the design file

5. A solid is constructed from two design files. Save and close your file (file name: *Nonlinear.des*).

Note that the original source solids (*Nonlinear1.des* and *Nonlinear2.des*) are not affected.

Advanced Model Editing Methods

Besides modifying the parameters of a solid part, you can edit a solid in several ways, including editing its faces and its body. These ways of modification are particularly useful for editing imported solid parts that do not have any parametric history.

Face Modification

Face modification includes deformation, removal, transformation, and offsetting.

Deformation

A selected face of a solid can be deformed by pulling it a specify distance as if the selected face is an elastic face. This is particularly useful for making a dome shaped object.

1. Start a new design file and set the measurement units to millimeters.

2. Double-click the Block from the Base Shapes tab of the palette to drop it to the base workplane.

3. Select face A indicated in Figure 5–12.

4. Select Feature > Modify Solids > Deform Face.

5. In the Deform Face dialog box, set height to 100 and select the OK button.

6. A face is deformed. (See Figure 5–13.) Save your file (file name: *Deform.des*).

7. Make a copy of the file by selecting File > Save Copy As and specifying a file name (*TransformFaces.des*).

8. Save one more copy of the file (file name: *OffsetFaces.des*).

9. Close the current *Deform.des* file.

*Figure 5–12
Face being
deformed*

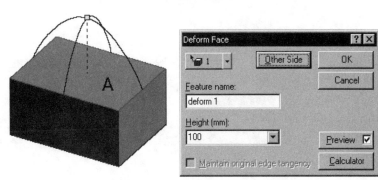

Figure 5–13
Selected face
deformed

Removal

By removing a set of selected faces, you can remove a feature, such as a hole or a pocket, from a solid. Naturally, you can delete the feature from the browser pane if the solid has a parametric history. Now perform the following steps to learn how to remove faces from a solid.

1. Start a new design file and set the measurement units to millimeters.

2. Double-click Block from the Base Shapes tab of the palette to drop it onto the base workplane.

3. Select face A indicated in Figure 5–14 and double-click Circular Pocket from the Pockets tab of the palette.

4. Select faces A and B indicated in Figure 5–15 and select Feature > Modify Solids > Remove Faces.

5. In the Remove Faces dialog box, select the OK button.

6. Save and close your file (file name: *RemoveFaces.des*).

Figure 5–14
Block dropped

Figure 5–15
Faces being
removed

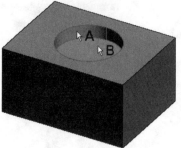

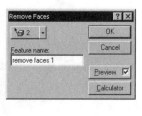

Transformation

To change the shape of a solid, you can translate selected faces a specified distance in a designated direction. Now perform the following steps to transform a face.

1. Open the file *TransformFaces.des*.

2. Select face A indicated in Figure 5–16 and select Feature > Modify Solids > Transform Faces.

3. In the Transform Faces dialog box, select the Transformation tab.

4. Set selection mode to Edges and select edge B indicated in Figure 5–16.

5. Select the OK button from the Transform Faces dialog box. (Note: You can also drag the yellow drag handle or specify some value in the distance field.)

6. A face is transformed. Save and close your file.

Figure 5–16 Face being transformed

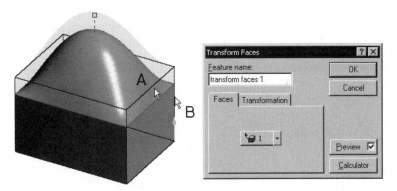

Offsetting

Offsetting is a special kind of face transformation in which selected faces are translated in a direction normal to the original face. Except for flat faces, the offset faces' profile and shape are changed. Now perform the following steps to offset a face of a solid.

1. Open the file *OffsetFaces.des*.

2. Select face A indicated in Figure 5–17 and select Feature > Modify Solids > Offset Faces.

3. In the Offset Faces dialog box, set the offset distance to 4 mm and select the OK button.

4. The selected face is offset. Compare the result with transformation.

5. Save and close your file.

Figure 5–17
Selected face being
offset

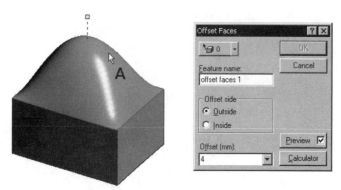

Body Modification

The entire solid body can be modified in several ways, including mirror, scale, and convert to a single feature.

Mirror

You can mirror the entire solid part, with the option of keeping or deleting the original solid. Now perform the following steps.

1. Start a new design file and set the measurement units to millimeters.

2. Double-click Step from the Base Shapes tab of the palette to drop it onto the base workplane.

3. Select Feature > Modify Solids > Mirror Solids.

4. In the Mirror Solids dialog box, set selection mode to Faces and select face A indicated in Figure 5–18. (Note: Prior to selecting face A, click in the free space to get rid of the drag handles after the Step has been dropped.)

5. Check the Keep original solids box and select the OK button.

6. The solid is mirrored and the original solid is kept in the final solid. Save your file (file name: *Body.des*).

Figure 5–18
Solid being mirrored

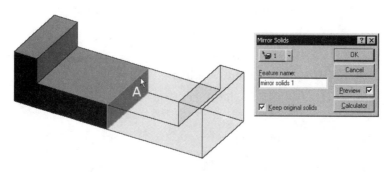

Scale

To resize the entire solid part as a whole, you can scale it uniformly by specifying a single scale factor or scale it in a non-uniform way by specifying the scale factors for width, depth, and height. Now perform the following steps:

1. Open the file *Body.des*, if you already closed it.

2. Select Feature > Modify Solids > Mirror Solids.

3. In the Scale Solid dialog box, select Uniform, set width scale to 0.8, and select the OK button.

4. Save your file.

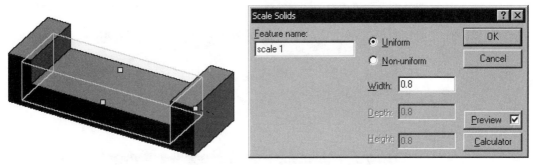

Figure 5–19 Solid being scaled uniformly

Convert to Single Feature

If you want to eliminate the parametric history of a solid part so that it can no longer be modifiable by manipulating its parameters, you convert all the features to a single feature.

1. Open the file *Body.des*, if you already closed it.

2. Select File > Save Copy As and specify a file name (*ScaleSolid.des*).

3. Set the browser pane to display the features in the file. (See Figure 5–20.)

4. Select Feature > Modify Solids > Convert to Single Feature.

5. In the warning message dialog box shown in Figure 5–21, select the OK button.

6. The features are converted to a single feature. (See Figure 5–22.) Save and close your file.

*Figure 5–20
Browser pane
showing the features
in the design file*

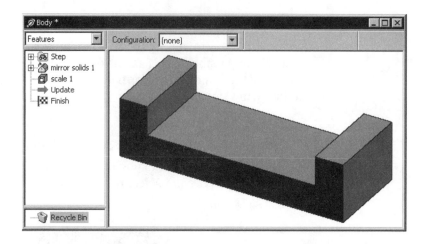

*Figure 5–21
Warning message*

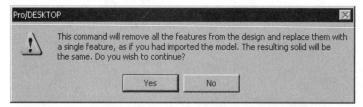

*Figure 5–22
Features converted
to a single feature*

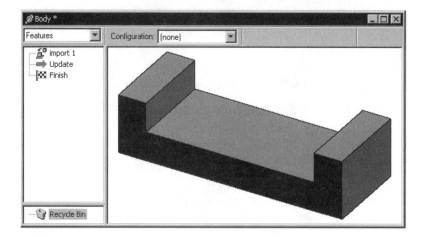

Treating Imported Surface Objects

Surface objects that are constructed by using some other kind of application and saved as IGES, STEP, or SAT format can be opened in Pro/DESKTOP. After opening, you can save them as Pro/DESKTOP surfaces. Although Pro/DESKTOP does not provide any tools for making surfaces, you can import surfaces constructed by using other computer-aided design applications and use them in two ways.

Replacement

You can replace a face or a number of faces of a solid with imported surfaces. Now perform the following steps.

1. Start a new design file and set the measurement units to millimeters.

2. Select File > Import > STEP file.

3. In the Import STEP File dialog box, select the Browse button, select the STEP file (*Surface.stp*) from the Chapter 5 folder of the CD accompanying this book, and select the OK button. A surface object is imported.

4. Double-click Cylinder from the Base Shapes tab of the palette. A cylinder is dropped.

5. Select Feature > Modify Solids > Replaces Faces.

6. Select face A indicated in Figure 5–18 to specify the face to be replaced.

7. Select the Replace By tab of the Replace Faces dialog box and select face B indicated in Figure 5–23.

8. Select the OK button. A face of the solid is replaced by an imported surface. (See Figure 5–24.) Save your file (file name: *ReplaceFaces.des*).

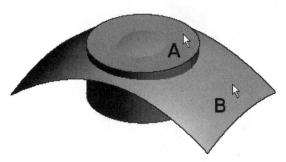
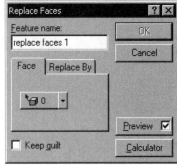

Figure 5–23 Face being replaced

Figure 5–24
Face replaced

Trim Solid

You can use a surface to trim a solid. After trimming, the trimmed face of the solid will resemble the shape of the surface used for trimming. Now perform the following steps.

1. Open the file *ReplaceFaces.des*, if you already closed it.

2. Select the Replace Faces icon from the features browser pane, right-click, and check the Suppress button.

3. Select Update from the Standard toolbar to update the change.

4. Select Feature > Modify Solids > Trim Solids.

5. Select surface A indicated in Figure 5–25. If the direction arrow is not the same as that shown in the figure, select Other Side from the Trim Solids dialog box.

6. Select the OK button.

7. The solid is trimmed. Note the difference between trimming and replacement. Save and close your file.

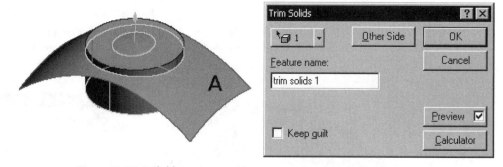

Figure 5–25 Solid being trimmed

Thicken Quilt

A quilt is a set of connected surfaces stitched together and treated as a single surface. You can thicken a surface or a quilt to produce a solid. Now perform the following steps.

1. Open the file *Surface.STP* from the Chapter 5 folder of the CD accompanying this book. (Note: You need to select STEP File from the Files of type pull-down list of the Open dialog box.)

2. Select Feature > Modify Solids > Thicken Quilt.

3. Select face A indicated in Figure 5–26.

4. In the Thicken Quilt dialog box, set thickness to 10 mm, check the Symmetric check box, and select the OK button.

5. The surface is thickened to form a solid. Save and close your file (file name: *Thicken.des*).

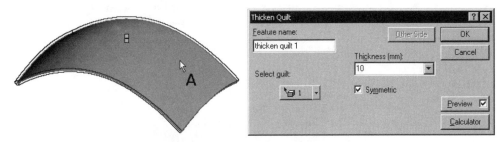

Figure 5–26 Surface being thickened

Miscellaneous Design Tools

The following paragraphs address various design tools available inside Pro/DESKTOP.

Variables

Variables are parameters in a file, including parameters used in sketching and making workplanes, parameters used in making features, results of measurements, and parameters controlling mating conditions. In addition, you can add user-defined variables. To manipulate variables, select Tools > Variables. Now perform the following steps to learn how to manipulate variables.

1. Start a new design file and set the measurement units to millimeters.

2. Double-click Block from the Base Shapes tab of the palette to drop it onto the base workplane.

3. Add dimensions to the sketch depicting the block. (See Figure 5–27.)

*Figure 5–27
Block dropped onto
the base workplane*

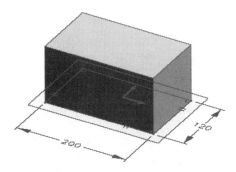

4. Select the upper face of the dropped block and drop a cylinder onto it.

5. Add dimensions in accordance with Figure 5–28.

Figure 5–28
Cylinder dropped
onto the upper face
of the block

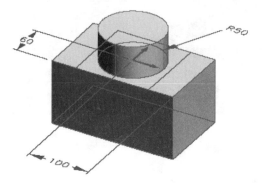

Now manipulate the variables.

6. Select Tools > Variables.

The Variables dialog box shown in Figure 5–29 has two panes and a button. The left pane is a browser delineating four kinds of variables organized in four folders: Workplanes, Features, Measurements, and Mating Conditions. The right pane is a table listing the variables related to workplanes, features, measurements, or mating conditions highlighted in the left pane. As for the Add button, it enables you to add a variable to selected workplanes, features, measurements, mating conditions, or to the design itself.

7. In the Variables dialog box, select Workplane 1 to show the variables related to this workplane.

8. Select Workplane 2 in the left pane of the Variable dialog box. This shows the variables related to Workplane 2.

Figure 5–29
Variables dialog
box showing
variables related to
workplane 1

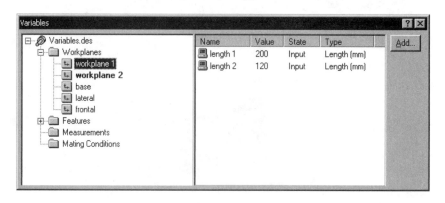

Figure 5–30
Variables related to
workplane 2

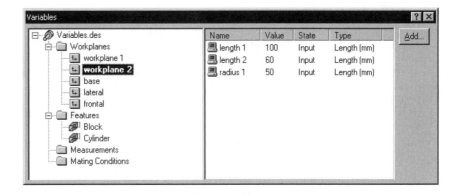

By comparing Figure 5–29 with Figure 5–30, you will find that there are length 1 and length 2 variables in both workplane 1 and workplane 2. To differentiate, you have to name a variable by specifying its full path. For example, the variable length 1 on workplane 1 should be specified as workplane 1\length 1, and variable 2 on workplane 2 should be called workplane 2\length 2.

Now you will add two user-defined variables.

1. Select Block from the Features section of from the left pane of the Variables dialog box.

2. Select the Add button.

3. In the New Variable dialog box, set variable name to height and variable value to 30.

4. Accept the default variable type and select the OK button. (Note: All system variables have computer icons. A user-defined variable has a face.)

5. Save your file (file name: *Variables.des*).

A user-defined variable is constructed.

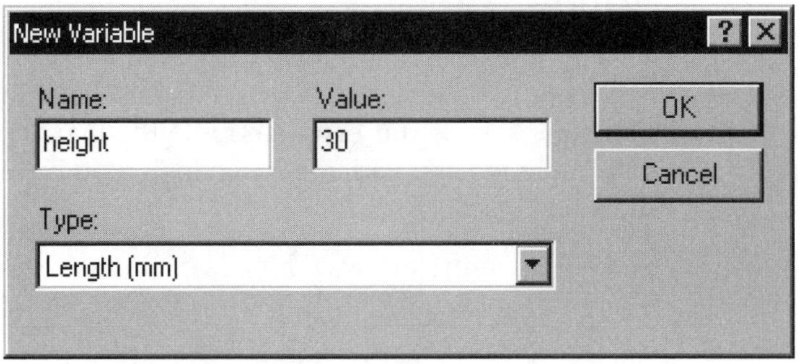

Figure 5–31
New Variable
dialog box

Design Rules

Design rules help establish parametric relationships between design variables. For example, you can set a dimension of a feature to be a function of a dimension of another feature of the solid part. Now perform the following steps to set up design rules in a design file.

1. Open the file *Variables.des*, if you already closed it.

2. Select Tools > Design Rules.

3. In the Design Rules dialog box shown in Figure 5–32, double-click next to the tick mark A. and then input the following expression. To use mathematical functions in your expression, select button B.

   *workplane 1\length 1 = 2 * workplane 2\length 1*

4. Select the tick mark next to the expression shown in Figure 5–32.

5. Input another expression as follows:

   Block\distance = Block\height

6. Select the Update button from the Standard toolbar and note the change in the solid part.

7. Save your file.

Figure 5–32
Design Rules dialog box

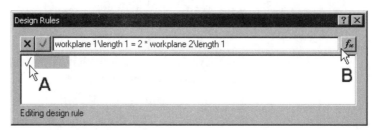

Solver

To optimize design variables in terms of other variables in a file, you can use the solver, which is accessible by selecting Tools > Solver. Using the solver, you can determine the maximum, minimum, or specific value for one variable by changing other variables. Now perform the following steps:

1. Open the file *Variables.des*, if you already closed it.

2. Select Tools > Solver.

3. In the Solver Parameters dialog box shown in Figure 5–33, click the Select button.

Figure 5–33
Solver Parameters
dialog box

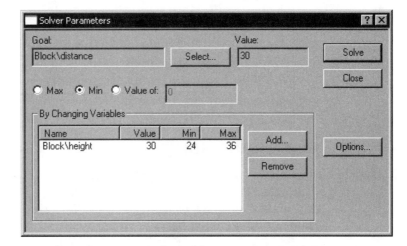

4. In the Set Variable dialog box shown in Figure 5–34, select the variable distance and select the OK button.

Figure 5–34
Set Variable dialog
box

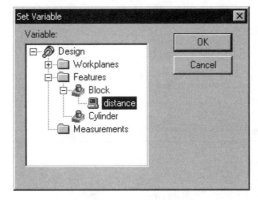

5. Select the Add button of the Solver Parameters dialog box.

6. In the Set Variable dialog box shown in Figure 5–35, select the variable height and select the OK button.

Figure 5–35
Set Variable dialog
box

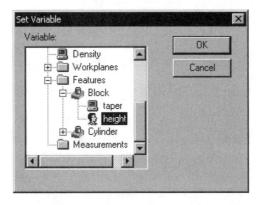

Now solve the variable to find out its minimum value.

7. Upon returning to the Solver Parameters dialog box, select the Min button.

8. Select the Solve button. The minimum value is evaluated.

9. Save your file.

Configuration

Configurations are sets of alternative parameters that you assign to variables in a file. By maintaining a set of configurations, you can evaluate alternative design scenarios. To set up configurations, select Tools > Configurations. (See Figure 5–36.)

Figure 5–36 Configurations dialog box

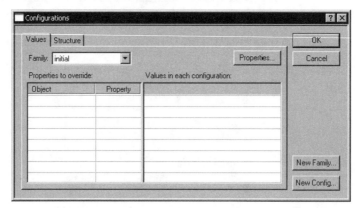

The Configurations dialog box has two tabs: Values and Structure. In the Values tab, you click the Properties button and select variables to put them in the Properties to override table. To set up alternative values to the variables, you select the New Config button to set up new configurations and assign values to each configuration listed in the Values in each configuration table. To set up different configurations using different sets of variables, select the New Family button to set up a number of families.

The second tab of the Configurations dialog box is the Structure tab. It lists the structure of the configurations in the form of a hierarchy.

Now perform the following steps to construct two configurations.

1. Open the file *Variables.des*, if you already closed it.

2. Select Tools > Configurations.

3. In the Configurations dialog box shown in Figure 5–36, select the Properties button in the Values tab.

4. In the Add/Remove Properties dialog box, select the variables workplane 1\length 1, workplane 1\length 2, workplane 2\radius 1, and workplane 2\length 2.

5. Select the ⇒ button to move the variables to the selected pane. (See Figure 5–37.)

6. Select the OK button,

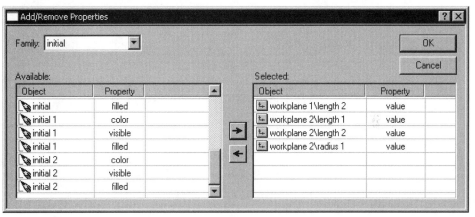

Figure 5–37 Add/Remove Properties dialog box

7. Upon returning to the Configurations dialog box, select the New Config button.

8. In the New Configuration dialog box, accept the default name and select the OK button.

Figure 5–38
New Configuration
dialog box

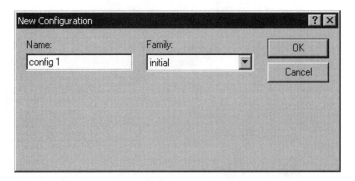

9. Repeat steps 7 and 8 to construct another configuration.

10. With reference to Figure 5–39, modify the variables in the Configurations dialog box. (Note: Values can be edited by clicking on them.)

Figure 5–39 Two configurations constructed

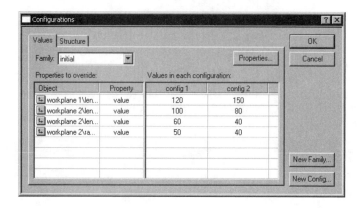

11. Select the OK button.

12. With reference to Figure 5–40, select each configuration one by one and select Update from the Standard toolbar to update the changes. Note the change in size of the component upon selecting different configurations.

13. Configurations are constructed. Save your file.

Figure 5–40 Configurations being updated

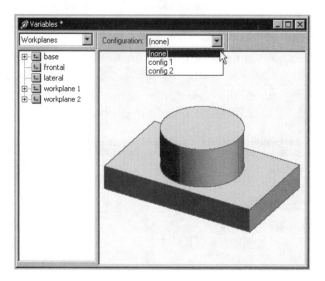

Animation

The animation tool enables you to study your design and collect data from transitioning through a series of configurations. Now perform the following steps.

1. Open the file *Variables.des*, if you already closed it.

2. Select Tools > Animation.

3. In the Keyframes tab of the Design Animation dialog box, click box A indicated in Figure 5–41.

4. Add two more rows to the table in accordance with Figure 5–42.

Figure 5–41
Design Animation
dialog box

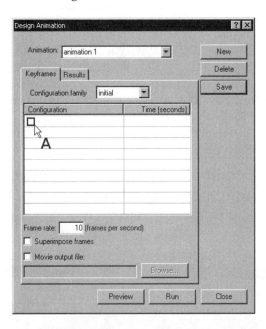

Figure 5–42
Keyframes added

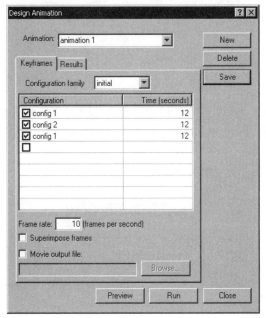

5. To output an AVI file, check the Movie output file box, specify a file name, and select the Run button.

6. Select the Save button and then the Close button.

Dimension Slider

The dimension slider enables you to dynamically modify the value of a variable to conduct a design study. To use the dimension slider, select Tools > Dimension Slider. After activating the Dimension Slider dialog box, you can select a variable in three ways: select a sketch dimension from the work plane, select a variable from the variable dialog box, or select a variable from the drop-down tree of the Variables box of the Dimension Slider. Now perform the following steps.

1. Open the file *Variables.des*, if you already closed it.

2. Select Tools > Dimension Slider.

3. Select the taper variable of the cylinder feature from the drop-down tree. (See Figure 5–43.)

Figure 5–43
Selecting a variable

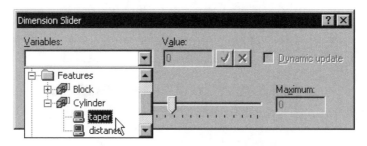

4. Check the Dynamic update box.

5. Move the slider bar to see the change in the model.

6. Set the taper angle to 20 and close the Dimension Slider dialog box.

7. Save your file.

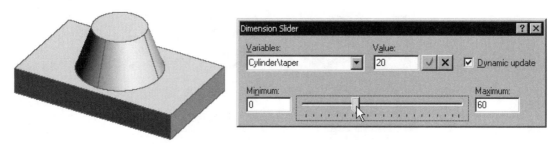

Figure 5–44 Dynamically updating the component

Measurement

To help evaluate the design as you work along in constructing the solid model, you can perform measurements as follows:

Dimension

You can measure the length, radius, diameter, closest distance, and angle between selected lines, workplanes, edges, and faces. Now perform the following steps.

1. Open the file *Variables.des*, if you already closed it.

2. Select Tools > New Measurement > Dimension.

3. Set selection mode to edges and select edges A and B indicated in Figure 5–45.

4. The result of measurement is shown in the dialog box. Select the OK button.

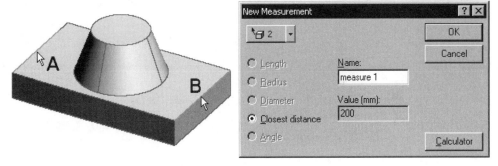

Figure 5–45 New Measurement dialog box

5. The result of measurement is listed in the Variables dialog box. (See Figure 5–46.)

6. To delete the measurement variable, select it from the Variables dialog box, right-click, and select Delete.

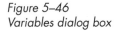

Figure 5–46
Variables dialog box

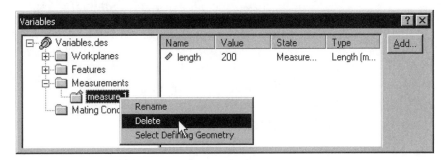

Sectional Properties

You can measure a sketch profile to generate a sectional properties report of the sketch. Now perform the following steps.

7. Select a sketch from the browser pane, right-click, and select Activate Sketch.

8. Select Tools > New Measurement > Sectional Properties. The result is shown in Figure 5–47, and the centroid is shown in the work pane.

Figure 5–47
Sectional Properties
dialog box

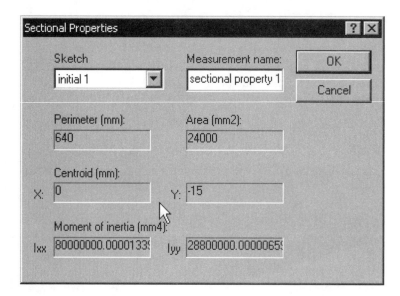

Mass Properties

You can measure the mass properties of a solid part. Information will be given in the Mass Properties dialog box, which has two tabs: Physical and Inertia. Mass, volume, surface area, and center of gravity will be shown in the Physical tab and moment of inertia, principal moments of inertia, principal axes, and radius of gyration will be shown in the Inertia tab.

9. Select Tools > New Measurement > Mass Properties. (See Figure 5–48.)

10. Select the Copy button to copy the result to the clipboard, where you can paste to other applications.

11. Close your file.

Figure 5–48
Mass Properties
dialog box

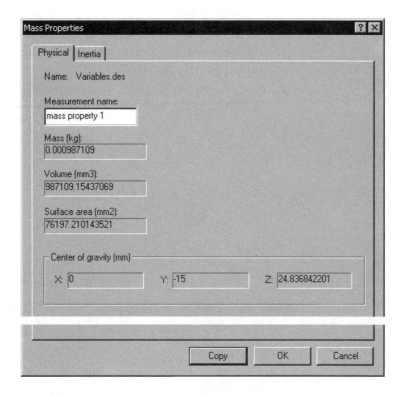

Exercises

Now work on the following exercises to enhance your knowledge about modeling.

Tray Project

You will construct the solid model of the component shown in Figure 5–49 by using the non-linear approach. Components for making this component can be found in the Chapter 5 folder of the CD accompanying this book.

Figure 5–49
Tray

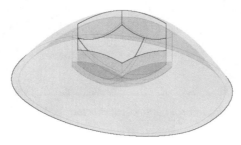

Now construct a solid part.

1. Open the file *Tray11.des* from the Chapter 5 folder of the CD accompanying this book.

2. Construct a loft solid feature through sketches A, B, and C indicated in Figure 5–50

Figure 5–50
Loft solid being
constructed

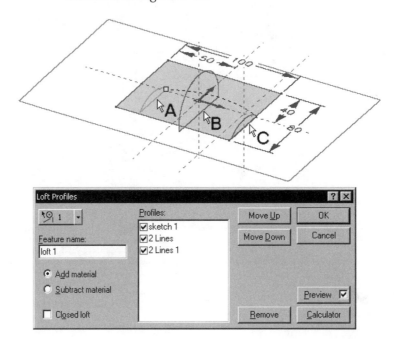

3. Activate the sketch residing on the Base workplane.

4. Extrude the sketch a distance of 50 mm to intersect the solid. (See Figure 5–51.)

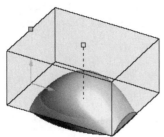 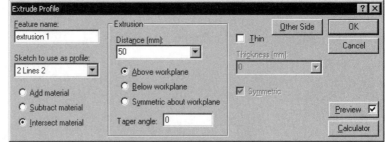

Figure 5–51 Extruded solid being constructed

5. The solid part is complete. (See Figure 5–52.) Save and close your file.

Figure 5–52
Solid part
constructed

Now construct the second solid part.

1. Open the file *Tray12.des* from the Chapter 5 folder of the CD accompanying this book.

2. Activate the sketch residing on the Frontal workplane

3. Revolve the sketch 360° about line A indicated in Figure 5–53.

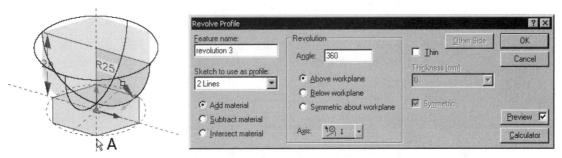

Figure 5–53 Sketch being revolved

4. Activate the sketch residing on the Base workplane and extrude the sketch a distance of 40 mm to intersect the solid. (See Figure 5–54.)

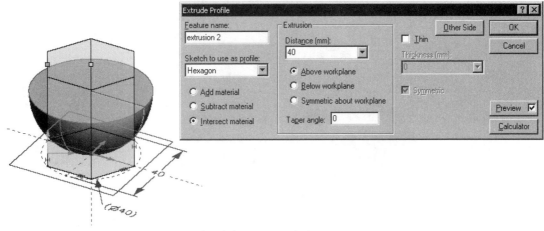

Figure 5–54 Sketch being extruded

5. The solid is complete. (See Figure 5–55.) Save and close your file.

Figure 5–55
Solid completed

Now construct a solid part from two components.

1. Start a new design file and set the measurement units to millimeters.

2. Add the components *Tray11.des* and *Tray12.des* to the file. (Alternatively, you can add the files *Tray01.des* and *Tray02.des* from the Chapter 5 folder of the CD accompanying this book.)

3. Select component A indicated in Figure 5–56 and select Feature > Use Component.

4. In the Use Component dialog box, select the OK button.

5. Select component B indicated in Figure 5–56 and select Feature > Use Component.

6. In the Use Component dialog box, check the Subtract material box and select the OK button.

7. A solid is formed. (See Figure 5–57.)

Figure 5–56
Components added

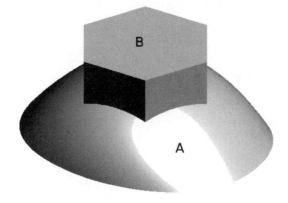

Figure 5–57
Components used
as features and
subtracted

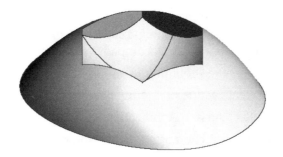

8. With reference to Figure 5–58, rotate the display and construct a shell feature.

9. In the shell feature, remove the bottom face, set shell thickness to 2 mm, and offset the faces outside.

10. The solid is complete. Save and close your file (file name: *Tray.des*).

Figure 5–58
Completed solid
viewed from
underneath

Joy Pad Project

You will construct a solid part and use imported surface features from STEP files constructed by using Rhinoceros.

1. Open the file *Joypad01.des* from the Chapter 5 folder of the CD accompanying this book.

2. Activate the sketch residing on the Base workplane, if it is not already activated.

3. Extrude the sketch a distance of 40 mm above the workplane with a taper angle of 2°. (See Figure 5–59.)

4. Select File > Import > STEP file.

5. In the Import STEP file dialog box, select the Browse button and then select the file *Joypad01.stp* from the Chapter 5 folder of the CD accompanying this book.

6. Select the OK button. A surface is imported. (See Figure 5–60.)

7. Select Feature > Modify Solids > Trim Solids.

8. Select surface A indicated in Figure 5–60.

9. If the direction arrow is not pointing upward, select the Other Side button from the Trim Solids dialog box.

10. Select the OK button. The solid is trimmed.

11. Activate the sketch residing on Workplane 1.

12. Extrude the sketch a distance of 50 mm above the workplane with a taper angle of 2°. (See Figure 5–61.)

Figure 5–59
Sketch being
extruded

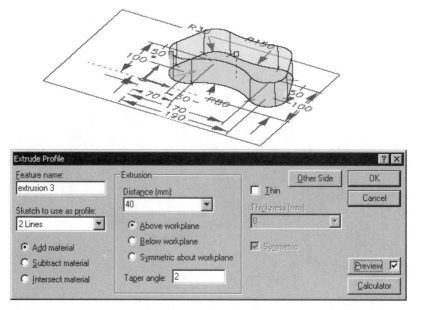

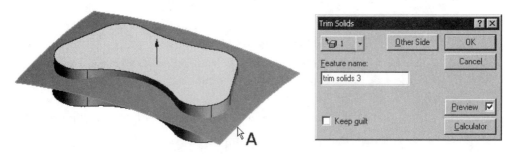

Figure 5–60 Surface imported and solid being trimmed

Figure 5–61
Sketch on workplane
1 being extruded

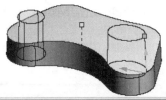

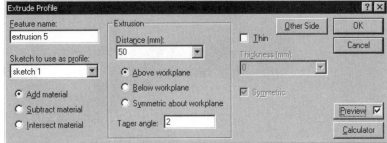

13. Import the STEP file *Joypad02.stp* from the Chapter 5 folder of the CD accompanying this book.

14. Use the imported surface to trim the solid, removing the upper portion of the solid. (See Figure 5–62.)

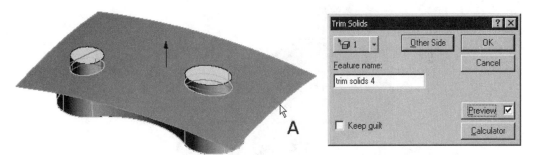

Figure 5–62 Surface imported and solid being trimmed

15. With reference to Figure 5–63, round off the edges with a radius of 2.5 mm.

Figure 5–63
Edges rounded

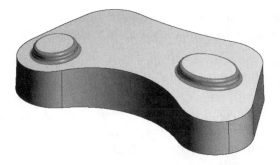

16. Round the edges highlighted in Figure 5–64 with a radius of 8 mm.

Figure 5–64
Edges being
rounded

17. Construct a shell feature with 2 mm offset outside. (See Figure 5–65.)

18. The solid is complete. Save and close your file.

Figure 5–65
Shell feature
constructed

Toy Plane Project

You will set up a number of configurations of the toy plane's assembly files and produce an animation of the configurations.

1. Open the file *PlaneS1.des* that you constructed in previous chapters.

2. With reference to Figure 5–66, rotate the display and hide the component Plane07 by selecting it from the browser pane, right-clicking, and checking the Hide box. This makes subsequent selection of faces easier.

Figure 5–66
Display rotated and
component Plane07
hidden

3. Select the component Plane03 and select Assembly > Fix Components. The component is fixed and will not move.

4. Select faces A and B indicated in Figure 5–67 and select Assembly > Orient.

5. In the Orient dialog box, set the angle to 0° and select the OK button.

Figure 5–67
Faces being
oriented

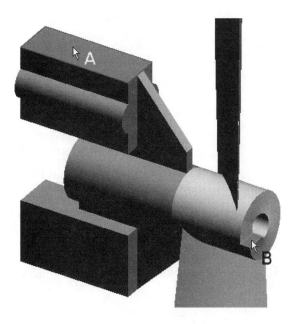

6. Unhide the component Plane07.

7. Select Tools > Configurations.

8. In the Configuration dialog box, select the Properties button and then choose the orient constraint that you just constructed.

9. On returning to the Configurations dialog box, set up four configurations, with values of 0, 45, 90, and 135. (See Figure 5–68.)

10. Select the OK button to close the Configurations dialog box.

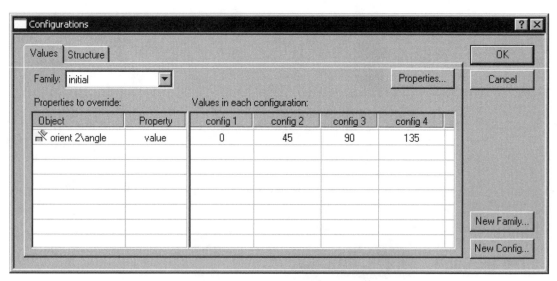

Figure 5–68 Configurations being constructed

11. Select Tools > Animation.

12. With reference to Figure 5–69, use the configurations config1, config 2, config 3, and config 4 in the animation's keyframe.

13. Check the Movie output file box and specify an AVI file name in the box below.

14. Select the Run button. (Note: A video compression dialog box appears. You must select a proper codec.)

15. Select the Save button to save the keyframe settings.

16. Select the Preview button to view how the animation will work.

17. Select the Close button.

18. Save and close the file.

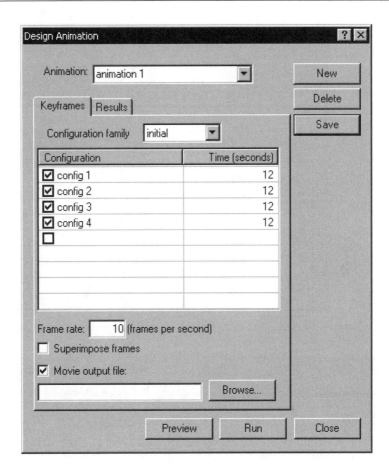

Figure 5–69
Design Animation
dialog box

Now you will work on the final assembly of the toy plane.

1. Open the file *Plane.des* that you constructed in Chapter 4.

2. Select Tools > Configurations.

3. Select the Properties button of the Configurations dialog box.

4. In the Add/Remove Properties dialog box, add PlaneS1 (1) Configuration and select the OK button. (See Figure 5–70.)

5. On returning to the Configurations dialog box, add four configurations and select the config1, config 2, config 3, and config 4 of PlaneS1 respectively, as shown in Figure 5–70.

6. Select the OK button to exit.

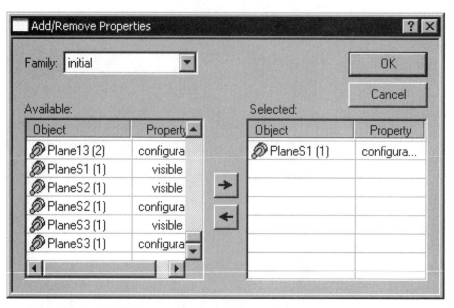

Figure 5–70 Add/Remove Properties dialog box

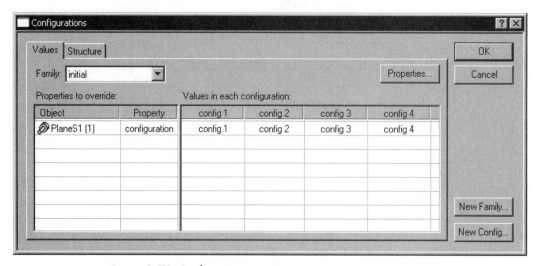

Figure 5–71 Configurations

7. Select Tools > Animation.

8. Similar to the settings made in the subassembly, use the configurations as keyframes, save an AVI file (*Plane.avi*), and save the animation. The dialog box is similar to Figure 5–69

9. The configurations and animation are complete. Save and close your file.

Summary

The basic way of constructing a solid part is to construct its features one by one and combine them as you construct them. Making the features in a linear way may impose certain limitations on the form and shape that can be produced. To widen the repertoire of form and shape, you may consider decomposing a complex object into two or more objects, constructing them in separate files, putting them together in a single file as if you are working on an assembly, applying assembly constraints, and using the components to build a complex solid. While building, you can add, subtract, and intersect components.

Besides the editing methods delineated in Chapter 2, you can edit a solid's face and body. Face modification includes deformation, removal, transformation, and offset. These methods of modifying a solid are particularly useful for modifying imported solid objects without parametric history. Body modification includes mirror, scale, and convert to a single feature. It must be noted that conversion of a solid to a single feature is irreversible. You should only use this command when you do not want the solid to be modified by downstream users of the model.

Although surface modeling tools are not available in Pro/DESKTOP, you can import surface objects from other applications, such as Rhinoceros, and use these surface objects in several ways. You can use a surface to trim a solid, you can replace selected faces of a solid by imported surfaces, and you can thicken a surface to obtain a solid.

To modify a solid part and evaluate various scenarios, you can manipulate variables, set up design rules, use the solver, establish configurations, run the animation tool, use the dimension slider, and make various kinds of measurements.

Review Questions

1. Explain the difference between the linear approach and the non-linear approach in solid modeling.

2. Describe how to construct a solid part in a non-linear way.

3. Delineate ways to modify the face and body of a solid part, in particular an imported solid part without parametric history.

4. How can imported surfaces be used in solid modeling?

5. Briefly explain the following terminology: variable, design rule, configuration, and animation.

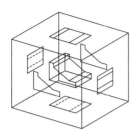

CHAPTER 6

Engineering Drafting

Objectives

This chapter outlines the key concepts of the orthographic projection system, delineates how to construct various kinds of engineering drawings from 3D solid models and assemblies of solid models, and explains how to add annotations to a drawing. After studying this chapter, you should be able to

❑ Explain the key concepts of orthographic projection

❑ Construct 2D engineering drawings from 3D solid parts and assemblies of solid parts

❑ Add annotations to an engineering drawing

Overview

Although computer models and related digital data are extensively used in most modern factories for downstream computerized manufacturing operations, there are still many occasions when you need to produce 2D orthographic drawings, in particular, for various kinds of processes that require a human being to interpret a design. That is why you need to learn how to produce 2D engineering drawings.

Constructing a 2D engineering drawing consists of two tasks: constructing 2D orthographic drawing views and adding annotations to the 2D engineering drawing views. If 3D solid models and virtual assemblies of solid models are already available, constructing 2D engineering drawing views can be performed in a semi-automatic way. You need only to start a drawing file, select a design file depicting a 3D solid model or an assembly of solid models, and let the computer generate related engineering drawing views. After the generation of drawing views, you complete the drawing by adding appropriate annotations.

Engineering Drafting Concepts

Engineering drafting is an engineering communication tool in which a 3D object is represented on a 2D drawing sheet through a set of orthographic views. The traditional way to construct an engineering drawing is to think about how a 3D object will look when you project it orthogonally on a 2D plane, and then to construct the orthographic views in accordance with your perception of the object's 2D appearance. You can construct the drawings manually or let the computer generate the drawing views for you if you have already constructed 3D computer models of the objects. After you construct the drawing views, you add annotations such as dimensions, text, geometric tolerances, surface finish symbols, welding symbols, and a parts list where appropriate.

Orthographic Projection

To depict a 3D object on a piece of 2D drawing paper, you use orthographic projection. The word "ortho" is a Greek word that means right or true. Orthographic projection is an engineering communication method to represent 3D objects on 2D drawing sheets by using multiple-view drawings. You project the 3D object perpendicularly onto a projection plane with parallel projectors. (See Figure 6–1.)

Figure 6–1
3D object projected onto an imaginery plane

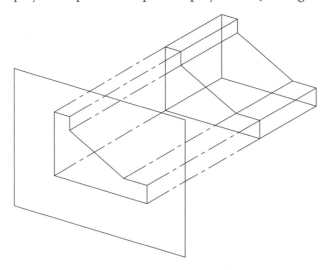

Basically, you can use six projection planes that are mutually perpendicular to each other to construct six drawing views showing the front, right side, left side, rear side, top, and bottom of the 3D object. Now you can imagine the 3D object inside the box and project views orthogonally onto the six walls of the box. (See Figure 6–2.)

Figure 6–2
Six projection
planes forming a
box with the 3D
object placed inside

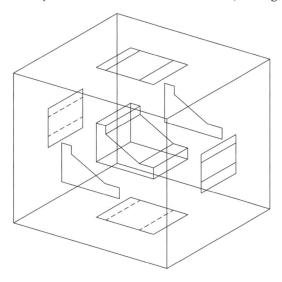

Because it is inconvenient to carry the box around, you cut and spread the box onto a common plane to obtain a drawing showing the six basic views. (See Figure 6–3.)

Figure 6–3
Cutting and
spreading the box

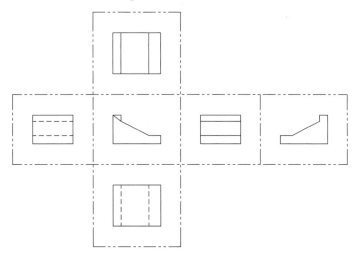

Projection Systems

There are two kinds of orthographic projection systems. In one system, you put the projection plane in front of the 3D object. In the other system, you place the projection plane at the far side of the 3D object. (See Figure 6–4 and compare it to Figure 6–1.)

Figure 6–4
Projection plane
placed at the far
side of the 3D
object

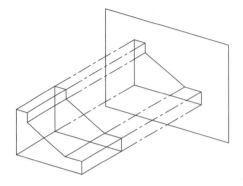

Similarly, there are six basic orthogonal views that form a box. (See Figure 6–5 and compare it to Figure 6–2.)

Figure 6–5
Six projection
planes

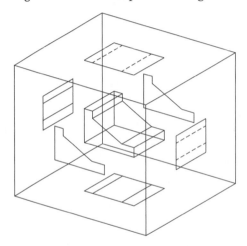

Again, you will cut and spread the box onto a common plane to get six basic views on a drawing sheet. (See Figure 6–6 and compare it to Figure 6–3.)

Figure 6–6
Six basic views

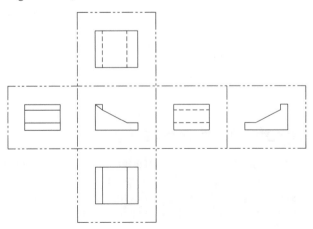

First and Third Angle Projection

In Figures 6–3 and 6–6, you can see that the front and rear views, the left and right side views, and the top and bottom views are quite similar. To describe this 3D object, three drawing views (front, side, and top) are sufficient. (See Figure 6–7.)

Figure 6–7
Three views of the
3D object in two
projection systems

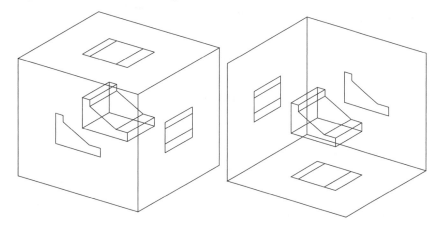

If you put the two projection systems together in 3D space, you will produce a very interesting picture. In Figure 6–8, one system falls neatly into the first quadrant, and the other into the third quadrant of the 3D space. Because we have to give the two projection systems names to identify which system we are using, we call one system the first angle projection system and the other the third angle projection system.

Figure 6–8
Two systems put
together

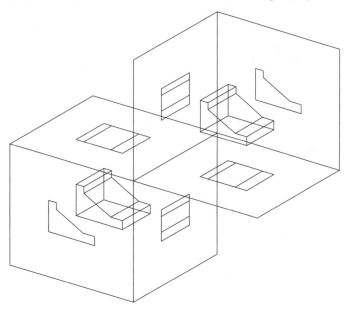

Projection Symbols

To indicate the system of projection that you are using, you place a symbol on your drawing sheet. The projection symbol is the engineering drawing of the front and side views of a conical object. (See Figure 6–9.)

Figure 6–9
Projection symbols

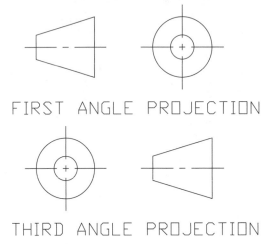

FIRST ANGLE PROJECTION

THIRD ANGLE PROJECTION

Generating 2D Drawings from 3D Models

Constructing a 2D engineering drawing begins with using a drawing template, details of which will be explained later, to start an engineering drawing file. It continues with linking the engineering drawing file to one or more design files, each depicting a solid part or an assembly of solid parts. You then specify a viewing direction and let the computer generate a 2D engineering drawing view from the object depicted in a linked design file. To complete the engineering drawing, you add annotations.

Drawing Standard

Because an engineering drawing is an engineering communication language, it is important that the drawing comply with appropriate national and international standards. Therefore, you need to set drafting standards before making a drawing. To comply with national or international standards, perform the following steps before starting an engineering drawing:

1. Close any engineering drawing file, if there are any opened drawing files.

2. Select Tools > Options.

3. In the Options dialog box, select the Drawing Standards tab, and select a standard from the Predefined standards list box.

4. Select the Modify button to make any change as may be necessary.

5. In the Standards dialog box, make necessary changes as may be necessary. For example, you can change the text height and leader length by manipulating the values specified in the Text and Leader tabs. (See Figure 6–10.)

Figure 6–10
Standards dialog box

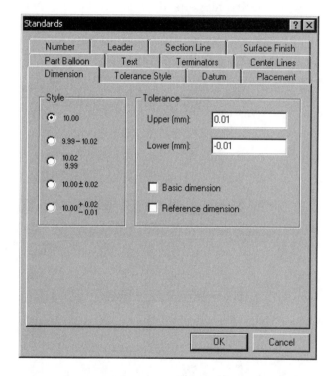

Using a Drawing Template and Selecting a Design File

By default, drawing templates are stored in the Drawing Format folder of the Pro/DESKTOP program folder. Now perform the following steps to use a template to start a new drawing file.

1. Select File > New > Engineering Drawing to open the New Engineering Drawing dialog box. (See Figure 6–11.)

2. In the New Engineering Drawing dialog box, you will find six template icons, depicting six different templates. Other than these templates, you can select the Browse button to display other templates available. Select the upper left icon.

3. At the lower portion of the dialog box, you will find three option buttons: New empty design, Design in session, and Open an existing design.

If you select New empty design and the OK button, a new design file will be opened together with a new engineering drawing file.

Here in Figure 6–11, the Design in session is greyed out because there is no design file being opened. If there are design files already open, you can select the Design in session button, select a design file from the list provided, and select the OK button to construct an engineering drawing of the selected design file.

Now select the Open an existing design button, select the Browse button, and select the file *Plane03.des* that you constructed earlier.

4. Select the OK button.

5. An engineering drawing is constructed. (See Figure 6–12.) Save your file (file name: *Plane03.dra*).

Figure 6–11
New Engineering
Drawing dialog box

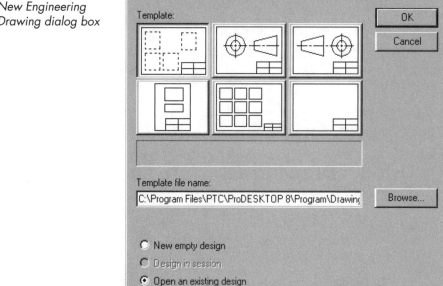

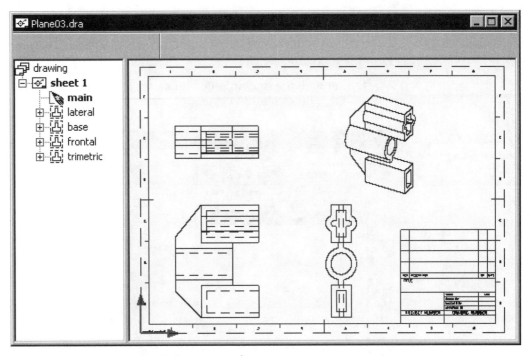

Figure 6–12 Engineering drawing

As can be seen, four drawing views, front view, top view, side view, and trimetric view, corresponding to the drawing views depicted in the drawing template, are constructed. In the browser pane shown in Figure 6–12, you will find a drawing sheet under which there are a number of icons representing a sketch and the drawing views.

Constructing Engineering Drawing Views

To construct an engineering drawing without pre-set drawing views, perform the following steps:

1. Select File > New > Engineering Drawing to open the New Engineering Drawing dialog box.

2. Select the lower right drawing template and select the OK button. An empty engineering drawing is started. To construct engineering drawing views for an empty engineering drawing, you need to have the related design file opened. If it is not open, open one.

3. Open the design file *Plane05.des*.

Drawing View from a Design File

To construct a drawing view from one of the opened design files, perform the following steps:

1. Activate the drawing file and select Drawing > Add Modeling View. (See Figure 6–13.)

Figure 6–13
Add Modeling View
dialog box

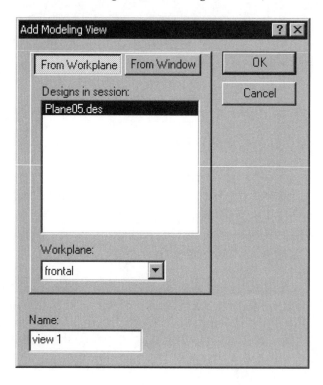

2. The Add Modeling View dialog box enables you to select a viewing direction for generating a drawing view. There are two buttons at the top of the dialog box: From Workplane and From Window. Select the From Workplane button, if it is not already selected.

3. Because there is only one design file open, the file in the Designs in session list is selected automatically. If there is more than one design file in the session, you have to select one.

4. In the Workplane pull-down list box, select frontal.

5. Select the OK button. A drawing view corresponding to the frontal workplane is constructed. (See Figure 6–14.)

Figure 6–14
Drawing view
constructed

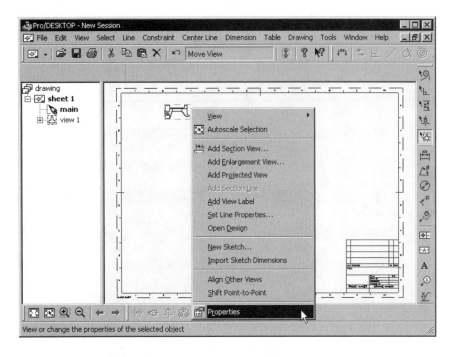

Isometric View

To construct an isometric view or a 3D view, perform the following steps:

1. Open the design file.

2. Manipulate the display so that the viewing window conforms to the required viewing direction of the drawing view.

3. Activate the drawing file and select Drawing > Add Modeling View.

4. Select the From Window button of the Add Modeling View dialog box and select the OK button.

Now change the drawing view's zoom scale, as follows:

1. Select the drawing view, right-click, and select Properties to modify the drawing scale, one of the properties of the drawing view.

2. In the Enlarge View area of the Drawing View tab of the Properties dialog box, check the Scale box and specify a zoom factor of 4. (See Figure 6–15.)

3. Select the OK button. The drawing view's zoom scale is changed. (See Figure 6–16.)

Projection Views

To construct a drawing view projected from an existing drawing view, perform the following steps.

1. Select the drawing view that you constructed and then select Drawing > Add Projected View.

2. In the Projection View dialog box shown in Figure 6–17, select Right and 3rd Angle and select the OK button. A projected view is constructed.

*Figure 6–15
Properties dialog
box*

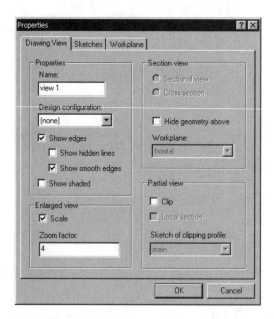

*Figure 6–16
Drawing view's
zoom scale changed*

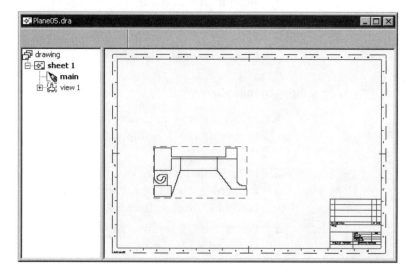

*Figure 6–17
Projection View
dialog box*

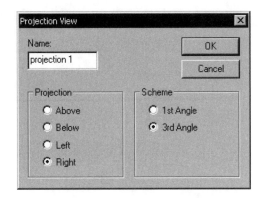

3. To move the drawing view, select it (View A shown in Figure 6–18) and drag it to a new location. In order to maintain alignment with its parent view, press the SHIFT key while dragging.

4. To align the drawing views, select view B shown in Figure 6–18 and select Drawing > Align other Views.

*Figure 6–18
Projected view
constructed and
being dragged to a
new position*

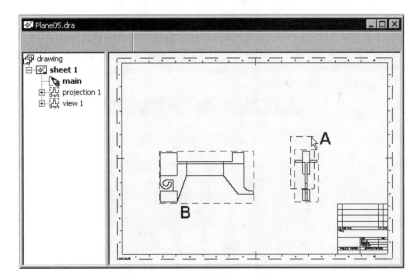

Section View

To add a section view, you have to construct a sketch with a line depicting the cutting plane and a construction line depicting the viewing direction. Now perform the following steps:

1. Using the line tools available from the Line menu, construct a straight horizontal line and a vertical construction line. (See Figure 6–19.)

Figure 6–19
Lines constructed

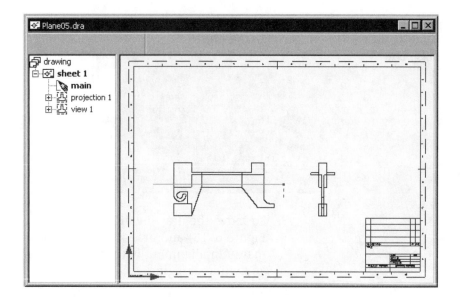

2. Select the front view and then select Drawing > Add Section View.

3. In the Section View dialog box shown in Figure 6–20, specify the section view's name and select the OK button. A section view is constructed.

Figure 6–20
Section View dialog box

4. With reference to Figure 6–21, select the section view and drag it to a new position. (Hold down the SHIFT key while dragging to maintain alignment between views.)

Figure 6–21
Section view
selected and
dragged

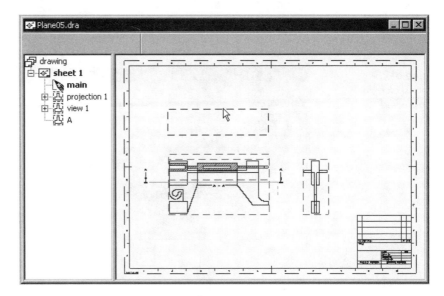

5. To align the views, if they are misaligned, select view
   A indicated in Figure 6–22 and select Drawing > Align
   Other Views.

Figure 6–22
Views being aligned

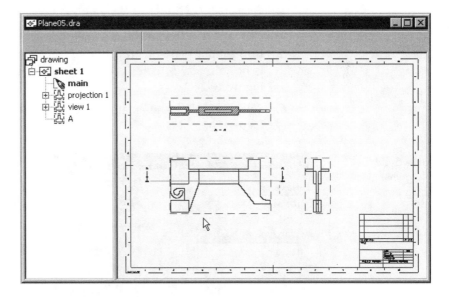

Enlargement View

To construct an enlarged view from an existing drawing view, perform
the following steps:

1. Select Sheet 1 from the browser pane, right-click, and select
   New Sketch.

2. With reference to Figure 6–23, construct a circle to specify the area of a drawing view to be enlarged.

Figure 6–23
New sketch
constructed

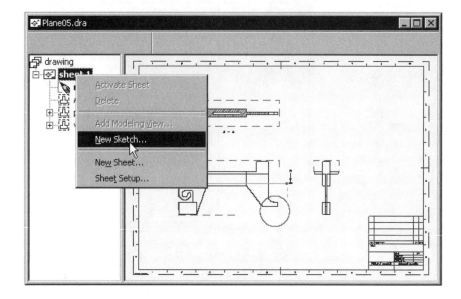

3. Set selection mode to Views and select the front view.

4. Select Drawing > Add Enlargement View.

5. In the Enlargement View dialog box, select sketch1, if it is not already selected, and select the OK button. (See Figure 6–24.)

6. Select and drag the enlarged view to location A indicated in Figure 6–25.

Figure 6–24
Enlargement View
dialog box

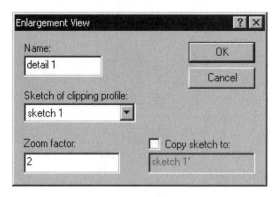

7. Select the sketch main, depicting the section plane, from the browser pane, right-click, and deselect Visible. (See Figure 6–26.)

Figure 6–25
View selected and dragged

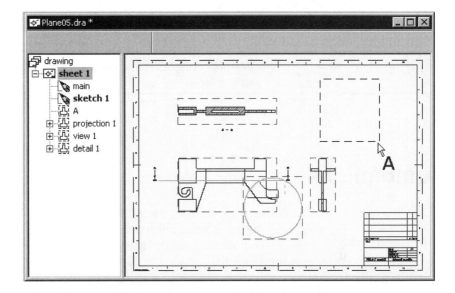

Figure 6–26
Sketch depicting the section plane made invisible

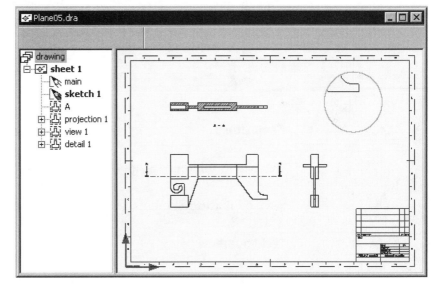

8. The drawing views are complete. Save and close the drawing file (file name: *Plane05.dra*).

Drawing Sheet

In a drawing, you can construct more than one drawing sheet on which you can generate different sets of engineering drawing views. For example, you can have a drawing sheet depicting an isometric view and a drawing sheet depicting the front, top, and side views.

Drawing File and Linked Design Files

Theoretically, you can link a number of design files to a single drawing file. For example, you can construct a drawing view of a design file together with a drawing view of another design file. However, you should not do so because that may cause confusion. In practical terms, a drawing file should link to a single design file.

Annotations

Annotations serve as supplements to the information provided by the drawing views. They are dimensions, note, centerlines, geometric tolerance symbols, surface texture symbols, weld symbols, parts list, and part reference balloons.

Dimensions	Use to depict the size of the object in drawings (for part drawings) and to depict distances between objects in drawings (for assembly drawings).
Notes	Use to provide a description in the drawing.
Centerlines	Use to illustrate axis and center locations.
Geometric Tolerance Symbols	Use to control the geometric shape of the object.
Surface Texture Symbols	Use to mandate surface finish requirement.
Weld Symbols	Use to illustrate how you will weld the component parts together in the assembly.
Parts List	Use to outline the particulars of the individual component parts.
Part Reference Balloons	Use in conjunction with the parts list to illustrate the locations of the individual parts in the assembly

Annotations in a Part Drawing

A drawing file for a 3D solid part is a full description of the 3D object. Along with the drawing views that show the shape and silhouettes of the object, you place dimensions, text, hole/thread note, hole table, leader text, centerlines, surface texture symbols, and geometric tolerance symbol.

Annotations in an Assembly Drawing

An assembly drawing is an engineering document describing how various component parts of the assembly are put together. Because you use part drawings to depict the 3D object, you do not need to repeat the dimensions, surface finish requirement, and geometric tolerances of the individual component parts in an assembly drawing. Basically, you need to have a parts list and a set of balloons included in the assembly drawing. In addition to the parts list and the set of balloons, you place dimensions, text, leader text, centerlines, surface texture symbols, geometric tolerance symbols, text, and weld symbols that are specific to the assembly.

Dimensions

Although dimensional information is already an integral part of the 3D part and assembly database, the dimensions of the objects are not readily perceivable if you do not display them explicitly on the drawing. To depict the actual size of a 3D solid in order to eliminate any possible errors that might arise in measuring the drawing, you add dimensions to your drawing.

Components of a Dimension

A dimension has four components:

Dimension Value	Indicates the actual size of the described feature.
Dimension Line	Specifies the direction of the described feature.
Extension Lines	Indicates the extents of the dimension line.
Geometric Tolerance Symbols	Projects from the feature to which the dimension refers.

In a dimension, there should be a small gap between the end of the extension line and the feature. The extension line should project a short distance away from the intersection of the dimension line. (See Figure 6–27.)

Figure 6–27
Components of a
dimension

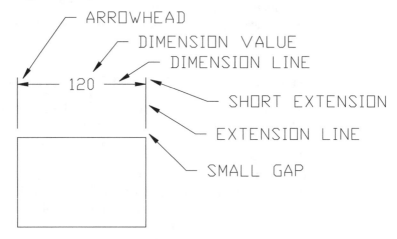

Dimensioning Principles

There are two basic principles to follow when you add dimensions to a drawing:

Appears Once	Each dimension required for the accurate definition of a feature should appear only once in the drawing. You should not assign more than one dimension to a feature.
No Calculation	As far as possible, you should not require the reader of your drawing to do calculation in order to obtain the dimension of a feature.

Because of the second principle, you might find it essential to add more than one dimension to a feature. In that case, you should put the additional dimension within parentheses to indicate an auxiliary reference dimension.

Now perform the following steps to add dimensions to a drawing.

1. Open the engineering drawing file *Plane03.dra*.

2. Select Dimension > Linear.

3. Select line A indicated in Figure 6–28.

4. Select line B and drag to location C indicated in Figure 6–28. A linear dimension is constructed.

Figure 6–28
Linear dimension
constructed

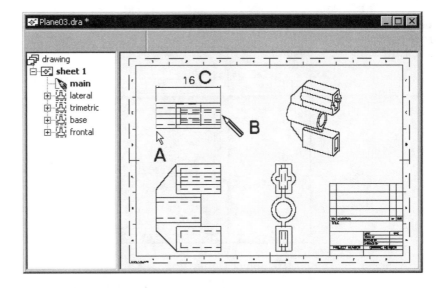

5. Select Dimension > Angular.

6. Select line A indicated in Figure 6–29.

7. Select line B and drag to location C indicated in Figure 6–29. An angular dimension is constructed.

Figure 6–29
Angular dimension
constructed

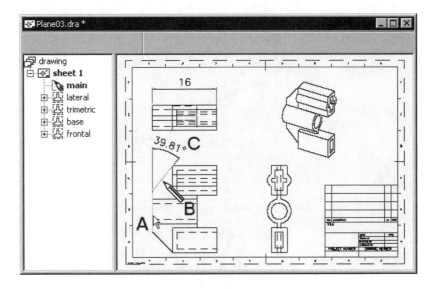

8. Select Dimension > Diametric.

9. Select circle A and drag to location B indicated in Figure 6–30. A diametric dimension is constructed. Note that a pair of centerlines is constructed automatically after you constructed a diametric dimension.

Figure 6–30
Diametric dimension
constructed

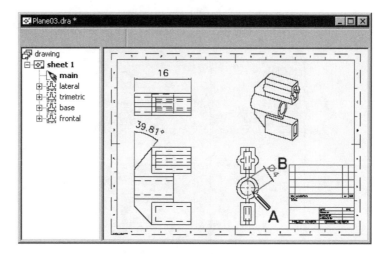

10. Select Dimension > Radial.

11. Select arc A and drag to location B indicated in Figure 6–31. A radial dimension is constructed.

12. Save your file.

Figure 6–31
Radial dimension
constructed

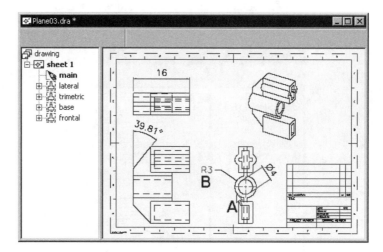

13. To retrieve dimensions you used in making the sketched solid features in an engineering drawing, set selection mode to Views, select a view, right-click, and select Import Sketch Dimensions.

14. Because the dimensions used in making the sketched solid feature may not be appropriate and may not conform to engineering standard practices, this way of adding dimensions, although fast, is not the best way of dimensioning. Now undo the last command.

*Figure 6–32
Importing
dimensions from
sketches*

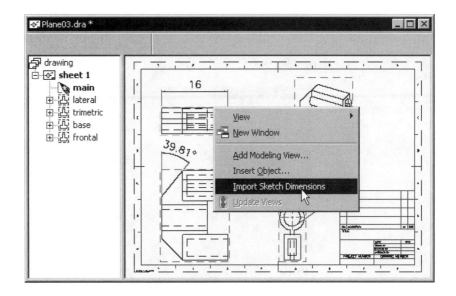

*Figure 6–32
Importing
dimensions from
sketches*

Notes

A note is a text string together with a leader. Now perform the following steps:

1. Open the file *Plane03.dra*, if you already closed it.

2. Select Dimension > Note.

3. Select line A and drag to location B indicated Figure 6–33.

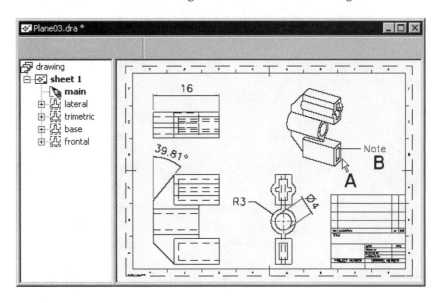

*Figure 6–33
Note being
constructed*

4. To set the text string content, double-click the text object.

5. In the Properties dialog box shown in Figure 6–34, fill in a single line text or multiple line text and select the OK button.

6. Save your file.

Figure 6–34
Properties dialog
box for text objects

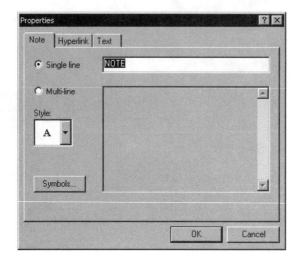

Centerlines

You use centerlines to indicate a center point and an axis of a cylindrical object. There are six ways to construct centerlines: common plane, common axis, mid-plane, pitch circle, center points, and phantom intersection, accessible from the Center Line menu.

Common Plane	Use to construct a centerline through two or more selected axial features that are lying in the same plane.
Common Axis	Use to construct a centerline through two or more selected coaxial features.
Mid-Plane	Use to construct a centerline midway between two selected planar features. Planar features do not necessarily have to be parallel.
Pitch Circle	Use to construct a circular centerline passing through the axes of selected features.
Center Points	Use to construct two centerlines and a center point for each selected axial object.
Phantom Intersection	Use to construct a pair of centerlines to depict an inferred intersection of two selected non-parallel planar features.

Now perform the following steps to add centerlines to an engineering drawing.

1. Open the file *Plane03.dra*, if you already closed it.

2. Set selection mode to Features and select edges A and B indicated in Figure 6–35.

3. Select Center Line > Phantom Intersection. Two centerlines depicting an inferred intersection are constructed.

Figure 6–35
Phantom intersection
centerlines being
constructed

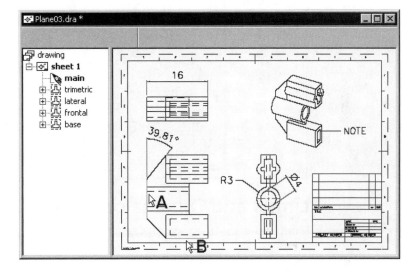

4. Select feature A indicated in Figure 6–36.

5. Select Center Line > Common Axis. A centerline passing through the axis is constructed.

Figure 6–36
Center line through
selected axis being
constructed

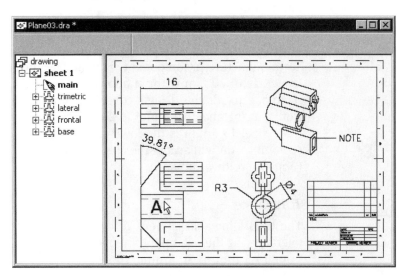

6. Select edges A and B indicated in Figure 6–37.

7. Select Center Line > Mid-Plane. A centerline midway between the selected edges is constructed.

8. Save your file.

Figure 6–37
Centerline midway between two edges being constructed

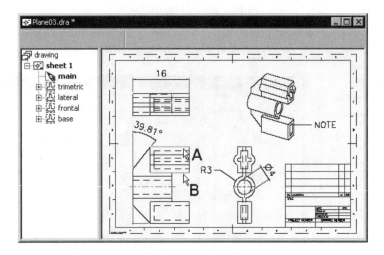

Limits and Dimension Tolerance

Limits refer to the range of size allowable on a component.

Basic Size

To design a component, we specify a size. The size we indicate in the drawing is the basic size. For example, if we are going to manufacture a circular shaft of diameter 20 mm, the 20 mm specification is the basic size.

Upper And Lower Limits

To manufacture a component, it would be impractical to insist on a fixed size of 20 mm without allowing any deviation. You should, with reference to all the design considerations, allow the size to deviate from, for instance, 19.5 mm to 20.5 mm. Here the 20 mm value is the basic size, 19.5 mm is the lower size limit, and 20.5 mm is the upper size limit. You will accept the component if it is made within this range (upper and lower limits).

Tolerance

The total allowable amount of deviation from the basic size is called tolerance. In the example of allowing the 20 mm basic size to deviate from 19.5 mm to 20.5 mm, you tolerate a size variation of 20.5 mm minus 19.5 mm. The tolerance has a direct impact on the cost of manufacture. The smaller the tolerance, the higher the manufacturing cost.

Now perform the following steps to specify a linear tolerance to a dimension.

1. Open the file *Plane03.dra*, if you already closed it.

2. Select dimension A indicated in Figure 6–38 and double-click.

3. In the Properties dialog box shown in Figure 6–39, set style and tolerance and select the OK button. The selected dimension is specified with a tolerance value.)

4. Save your file.

Figure 6–38
Linear tolerance
specified

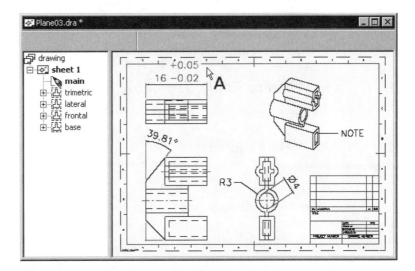

Figure 6–39
Properties dialog
box

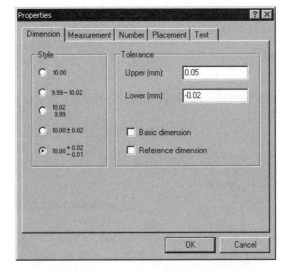

Geometric Tolerance

You can copy and paste feature control frames, surface texture symbols, datum identifiers, datum targets, weld notes, and user-defined symbols, with and without leaders. This way, you can save a lot of time in inserting a number of identical symbols in a drawing. Geometric tolerance concerns the maximum permissible overall variation of form or position of a feature. A geometric tolerance defines a tolerance zone within which features of a component are to be contained. Depending on the nature of the geometric tolerance, the tolerance zone can be the area within a circle, the area between two concentric circles, the area between two equidistant lines, the area between two parallel lines, the space within a cylinder, the space between two concentric cylinders, the space between two equidistant planes, the space between two parallel planes, or the space within a parallelepiped.

When to Apply Geometric Tolerance

There are four general conditions when geometric tolerances are required:

Form Control	You apply geometric tolerance when the application of dimension tolerances alone does not impose the desired control over the form and shape of a component. After you apply geometric tolerance to a feature, it will take precedence over the form control imposed by the size tolerance.
Machinery and Techniques	You apply geometric tolerance to mandate the use of appropriate machine tools and techniques to produce the product.
Remote Manufacture	You specify geometric tolerance in the engineering drawing to provide full information on the functional requirement of the product.
Features without Strict Size Control	You apply geometric tolerance to components without strict size control, such as to control the flatness of a surface table.

General Principles

There are five general geometric tolerance principles, concerning scope, datum, tolerance value, feature shape, and form control.

Scope	You apply geometric tolerance to the whole length or surface of a feature, unless otherwise specified.
Datum	In choosing a reference datum for a geometric tolerance, you select a datum with adequate accuracy, and you consider the functionality of the part.
Tolerance Value	You specify geometric tolerance value relative to feature size, unless otherwise stated.
Feature Shape	The final feature shape can take any form within the tolerance zone if no further control is given.
Form Control	Some geometric tolerance can automatically control other kinds of form errors.

Symbol

You use geometric tolerance symbols to detail the tolerance applied to the geometric shape of the 3D object. The main body of a geometric tolerance is a feature control frame. In the feature control frame, there are two or more compartments: In the first compartment, you put a geometric tolerance symbol depicting the kind of control you are imposing on the geometry. In the second compartment, you put the geometric tolerance value that defines a geometric tolerance zone. The third and fourth compartments are optional; you can add the datum reference name(s). When you specify datum reference(s) in the geometric tolerance symbol, you specify a datum by using a datum identifier or a datum target. A datum identifier specifies the entire face of the indicated feature as datum reference. A datum target specifies a zone of the face of a feature as datum reference. Figure 6–40 shows a feature control frame controlling the position of a feature to fall within a circular tolerance zone of 0.05 mm diameter with reference to datum A and datum B. In the feature control frame, you can, as with dimensioning, include dual dimensions.

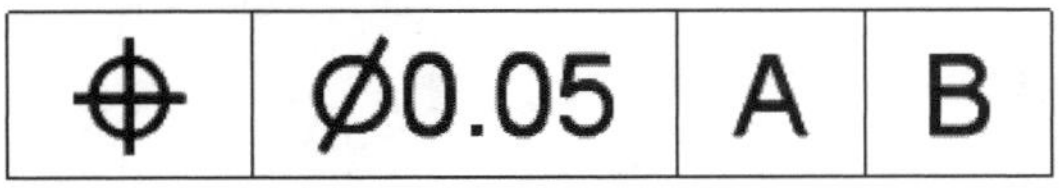

Figure 6–40 Feature control frame with four compartments

$$\oplus \quad \boxed{\varnothing 0.05} \quad A \quad B$$

Figure 6–41 shows various kinds of geometric characteristic symbols that you can insert in the first compartment of the feature control frame.

Figure 6–41
Geometric
characteristics

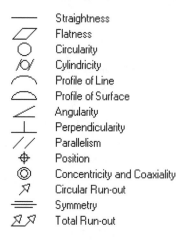

——	Straightness
▱	Flatness
○	Circularity
/○/	Cylindricity
⌒	Profile of Line
◠	Profile of Surface
∠	Angularity
⊥	Perpendicularity
//	Parallelism
⊕	Position
◎	Concentricity and Coaxiality
↗	Circular Run-out
≡	Symmetry
↗↗	Total Run-out

Now perform the following steps to add a geometric tolerance to your drawing.

1. Open the file *Plane03.dra*, if you already closed it.

2. Select Dimension > Datum Feature.

3. Select edge A and drag to location B indicated in Figure 6–42.

4. Double-click text object B in Figure 6–42.

5. In the Properties dialog box, set the datum name to A, select Datum label style, and select the OK button. (See Figure 6–43.) A datum feature is constructed.

Figure 6–42
Datum feature being
constructed

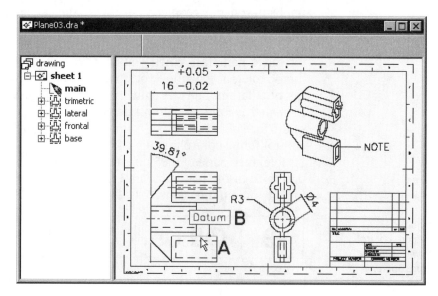

Figure 6–43
Properties dialog
box

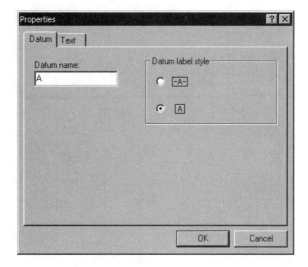

6. Select Dimension > Geometric Tolerance.

7. Select edge A and drag to location B indicated in Figure 6–44

8. Double-click the feature control frame B in Figure 6–44.

9. Select Parallelism from the Feature Control pull-down list, set tolerance to 0.1 mm, specify datum A, and select the OK button. (See Figure 6–45.)

10. A feature control feature governing the parallelism of a feature in relation to another feature is constructed. (See Figure 6–46.) Save your file.

Figure 6–44
Feature control
frame being
constructed

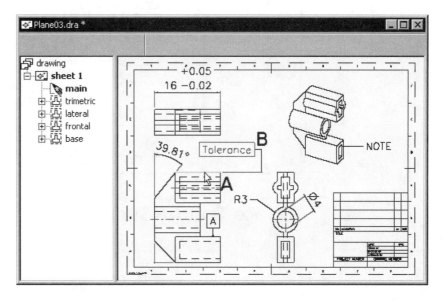

*Figure 6–45
Properties dialog
box*

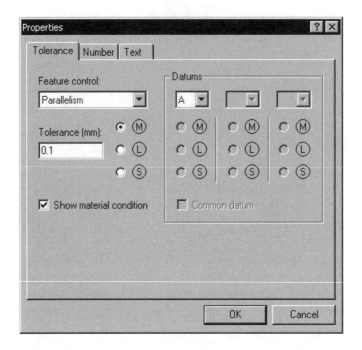

*Figure 6–46
Feature control
frame constructed*

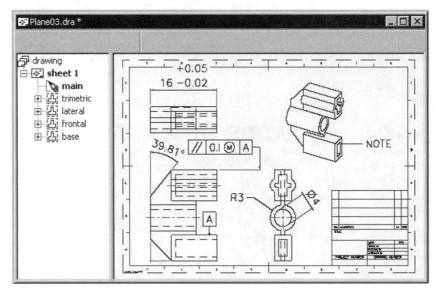

Surface Texture Symbols

To control the surface finish of a component, you specify surface finish
symbols in your engineering drawing.

Basic Symbol

The basic form of a surface finish symbol is similar to the letter "v." In addition to the basic symbol, there are two derivatives: material removal required and material removal prohibited. (See Figure 6–47.)

*Figure 6–47
Surface finish
symbols (from left to
right): basic, material
removal required,
and material
removal prohibited*

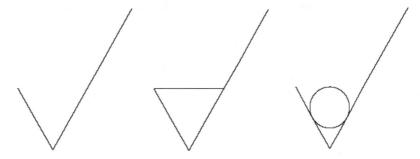

With the basic symbol, machining of the component is optional, as long as the required roughness value is achieved. With the material removal required symbol, machining is mandatory. With the material removal prohibited symbol, machining is not allowed, but the required roughness value must be achieved.

Roughness Value

It is pointless to specify a surface finish symbol without stipulating the roughness value of the surface. You can simply state the maximum roughness or a range of roughness. Figure 6–48 shows a roughness value of 0.8 micrometers, and Figure 6–49 shows a symbol mandating a roughness value between 0.8 and 0.4 micrometers.

*Figure 6–48
Roughness value
specified*

*Figure 6–49
Maximum and
minimum roughness
values specified*

Direction of Lay

Direction of lay refers to the machining mark or pattern left on the surface after it is machined. For example, if you use sandpaper to polish a surface and rub the sandpaper in a linear direction along the surface, the direction of lay on the surface is linear and parallel to the direction of your hand's movement. There are seven directions. (See Figure 6–50.)

Figure 6–50
Lay

$=$	Parallel to Plane of Projection
$\perp$	Perpendicular to Plane of Projection
X	Crossed in Two Slant Directions
M	Multidirectional
C	Circular Relative to Center
R	Radial Relative to Center
P	Particulate, Nondirectional

A symbol indicating a lay parallel to the plane of projection of the drawing view is shown in Figure 6–51. Specification of lay is optional.

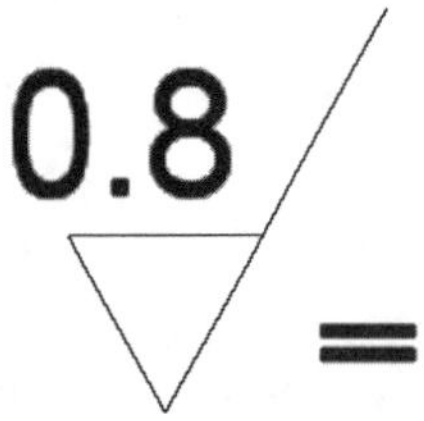

Figure 6–51
Direction of lay
parallel to the
projection of the
drawing view

0.8

Machining Allowance

If a component has to be machined to achieve the required surface finish, you add an allowance on the component. To specify the machining allowance, you state it in the surface finish symbol. A surface finish symbol depicting a machining allowance of 2 mm is indicated in Figure 6–52.

Figure 6–52
Machining
allowance specified

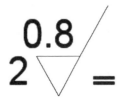

0.8
2

Production Method

To stipulate the way to achieve the required surface finish, you specify a production method. Figure 6–53 shows a surface finish symbol with production method specified. Specification of the production method is optional.

Figure 6–53
Production method
specified

Sampling Length

To specify the length for taking a sample of the surface for roughness average value measurement, you specify the sample length. (See Figure 6–54.) As with the production method, specification of the sample length is optional.

Figure 6–54
Sample length
specified

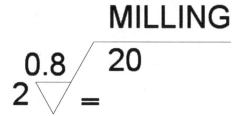

Now perform the following steps to construct a surface finish symbol.

1. Open the file *Plane03.dra*, if you already closed it.

2. Select Dimension > Surface Finish.

3. Select edge A and drag to location B indicated in Figure 6–55.

Figure 6–55
Surface texture
symbol being
constructed

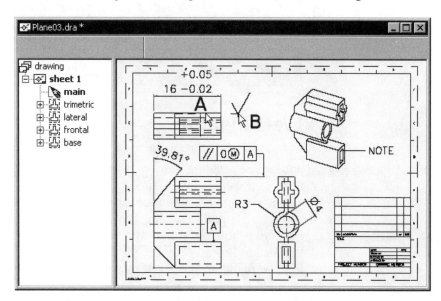

4. Double-click symbol B indicated in Figure 6–55.

5. In the Properties dialog box, select By any production method, specify INJECTION MOLD as production method, set roughness value to 0.4, and select the OK button.

6. A surface texture symbol is constructed. Save your file.

Figure 6–56
Properties dialog
box

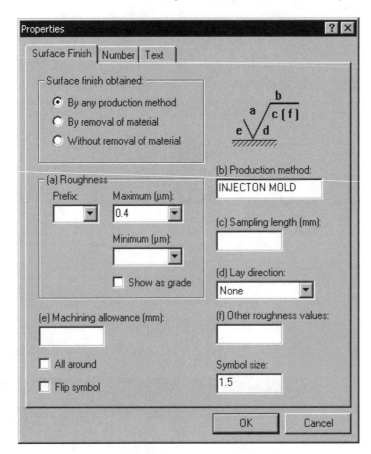

Weld Symbols

To specify how two or more components are to be welded together, you add a welding symbol to your engineering drawing. The basic form of a weld symbol consists of a horizontal line and a leader line. The leader indicates the location of the weld, and the horizontal line carries a symbol depicting the kind of weld. If the symbol is placed below the horizontal line, the weld is to be made on the near side of the leader. If the symbol is placed above the horizontal line, the weld is to be made on the other side of the leader. Figure 6–57 shows a fillet weld to be made at the near side of the leader. The symbols that you specify on the horizontal line depicting various kinds of weld are shown in Figure 6–58.

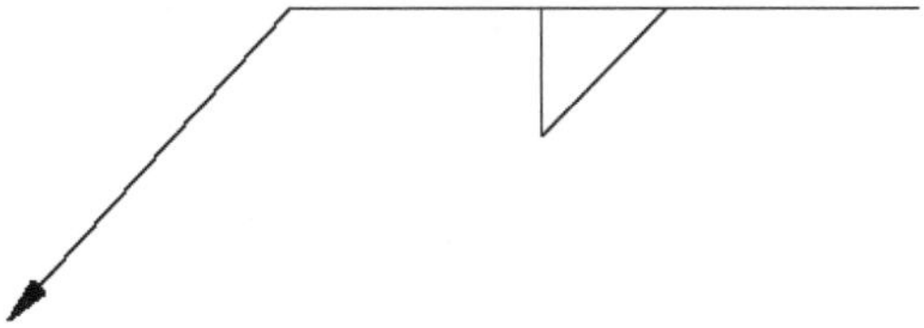

Figure 6–57 Weld symbol depicting a fillet weld

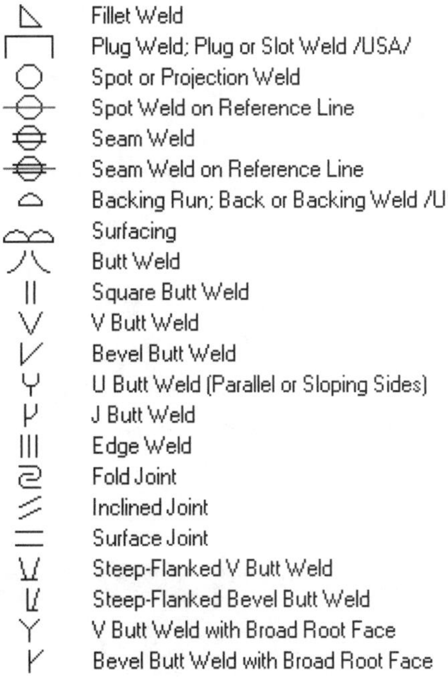
Fillet Weld
Plug Weld; Plug or Slot Weld /USA/
Spot or Projection Weld
Spot Weld on Reference Line
Seam Weld
Seam Weld on Reference Line
Backing Run; Back or Backing Weld /USA/
Surfacing
Butt Weld
Square Butt Weld
V Butt Weld
Bevel Butt Weld
U Butt Weld (Parallel or Sloping Sides)
J Butt Weld
Edge Weld
Fold Joint
Inclined Joint
Surface Joint
Steep-Flanked V Butt Weld
Steep-Flanked Bevel Butt Weld
V Butt Weld with Broad Root Face
Bevel Butt Weld with Broad Root Face

Figure 6–58 Symbols for various kinds of welds

Field Weld

If a welding joint has to be carried out on site, you specify a field weld indicator. The field weld symbol is optional. (See Figure 6–59.)

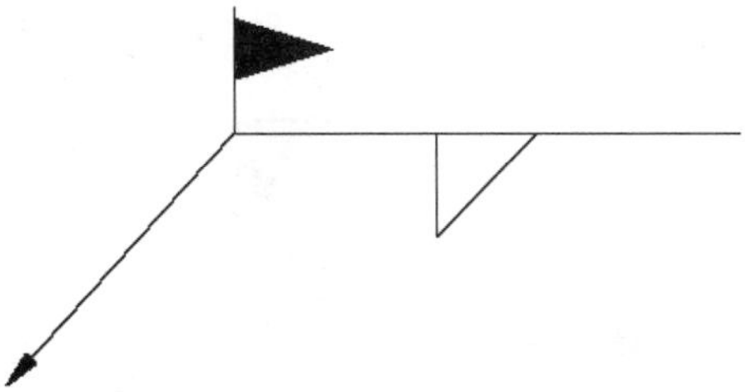

Figure 6–59 Field weld indicator added to the weld symbol

All Round Welding

If a joint is to be carried out all round an object, you specify an all round indicator instead of adding a set of symbols around the object. (See Figure 6–60.)

*Figure 6–60
All round indicator
added to the weld
symbol*

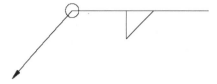

Notes

If additional information is required, you add notes to the weld symbol. (See Figure 6–61.)

*Figure 6–61
Notes included in a
weld symbol*

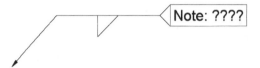

Note: ????

Now perform the following steps to add a weld symbol to a drawing.

1. Open the drawing file *Weld.dra* from the Chapter 6 folder of the CD accompanying this book.

2. Select Dimension > Welding Symbol and then select edge A and drag to location B indicated in Figure 6–62.

*Figure 6–62
Weld symbol being
constructed*

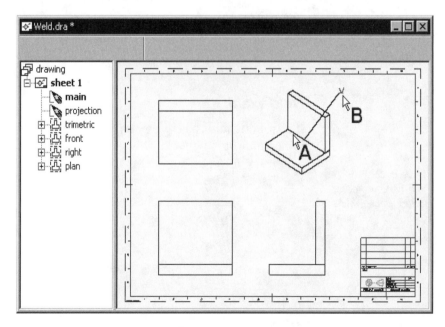

3. Double-click weld symbol B indicated in Figure 6–62.

4. With reference to Figure 6–63, specify details of the weld joint and select the OK button.

5. A weld symbol is constructed. Save your file.

Figure 6–63
Properties dialog
box

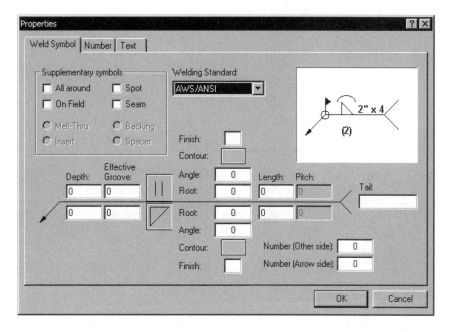

Parts List

It is standard engineering practice to include a parts list in an assembly drawing. A parts list provides information about the quantity and references the parts of the assembly. Now perform the following steps to add a parts list to a drawing depicting the assembly of components.

1. Open the drawing file *Weld.dra*, if you already closed it.

2. Select Table > Add Parts List.

3. In the Add Parts List dialog box, select the OK button.

4. Select and drag the parts list to location A indicated in Figure 6–64.

5. A parts list is constructed. Save your file.

Figure 6–64
Parts list added

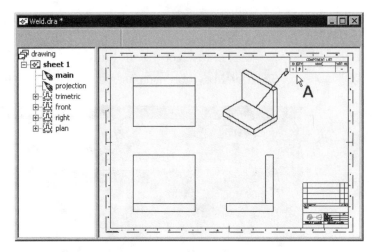

To modify the attributes in the parts list, which refer to the individual components, perform the following steps:

1. Open the design file *Weld01.des* from the Chapter 6 folder of the CD accompanying this book.

2. Select File > Properties.

3. In the Properties dialog box shown in Figure 6–65, specify Part Number and Part description.

4. Save and close your file.

When you return to the engineering drawing file, information specified in the design file's Properties dialog box will be listed in the parts list.

Figure 6–65
Properties dialog box

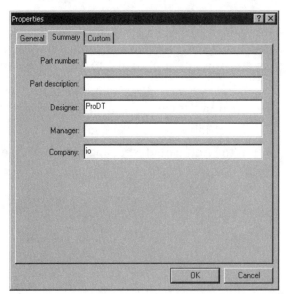

Part Reference Balloons

To reference the individual parts in the drawing view and the parts list, you use a special kind of leader: balloons. Now perform the following steps to add balloons to a drawing depicting an assembly.

1. Open the drawing file *Weld.dra,* if you already closed it.

2. Select Dimension > Part Reference Balloon.

3. Select A and drag to B indicated in Figure 6–66.

4. A balloon is constructed. Save and close your file.

Figure 6–66
Balloon being
constructed

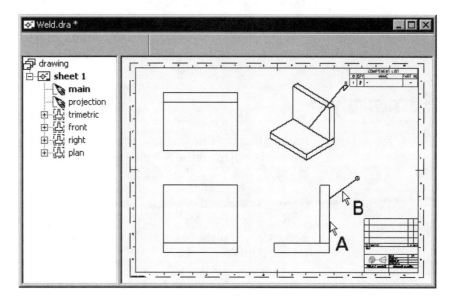

Templates

Templates are engineering drawing files with backgrounds and layouts that can be reused for multiple drawings. Typically, a template consists of a set of border lines, a sheet header block, and related drawing information.

The drawing templates that you use to construct an engineering drawing are saved in the Drawing Formats folder of the Pro/DESKTOP program folder. You can open these files to modify them to suit your needs.

Exercises

Now enhance your knowledge by working on the following exercises.

Ballpoint Pen Project

Construct engineering drawings for the ballpoint pen assembly and its components.

Toy Plane Project

Construct engineering drawings for the toy plane assembly and its components.

Summary

An engineering drawing is a set of orthographic drawing views depicting a 3D object on a 2D engineering drawing sheet. To construct a drawing, you use a drawing file.

Using a drawing file, you construct an engineering drawing associated with one or more design files. Although linking to multiple design files is allowed, you should only link a drawing file to a single design file to avoid confusion.

Constructing a drawing consists of two major tasks: constructing drawing views depicting the linked design file and constructing annotations. Annotations include dimensions, notes, centerlines, geometric tolerance symbols, surface texture symbols, weld symbols, parts list, and part reference balloons.

Review Questions

1. What are the three kinds of Pro/DESKTOP files? Which one will you use to construct an engineering drawing and how?

2. How many kinds of drawing views can you construct in a drawing file? Use simple sketches to illustrate your answer.

3. What kind of annotations will you add to an engineering drawing depicting a solid part, a sheet metal part, and an assembly of parts?

 **CHAPTER 7**

Photo Album

Objectives

This chapter delineates the ways to construct rendered images and animations of solid parts and assemblies of solid parts. After studying this chapter, you should be able to

❐ Construct rendered images and animations of solid parts and assemblies of solid parts stored in a design file

Overview

Besides engineering drawings to facilitate certain kind of manufacturing processes requiring 2D orthographic drawings, photo-realistic images and animations of the computer models may be required in presentation and visualization. To make an image of a 3D solid or assembly of 3D solids realistic, material properties, background, camera settings, and lighting need to be taken into consideration. If an animation of configurations is already constructed in the related design file, you can also output a photo-realistic animation.

Functions of Photo Album

Pro/DESKTOP's photo album is a visualization tool, enabling you to apply material properties to selected faces and components of solid parts, set lighting in the environment, superimpose the solid part over a background image, set camera viewing options, and output rendered images and animations. Figure 7–1 shows the rendered image of the toy plane.

Figure 7–1
Rendered image

Constructing Rendered Images

Constructing rendered images requires a photo album file. Because the geometry of the solid parts is already defined in design files, a photo album file's geometric objects are referenced to linked design files. Similar to a drawing file, a photo album file can be linked to multiple design files. However, for easy reference, it is recommended that a photo album file be referenced to a single design file.

In a photo album file, you can construct a number of images, each depicting a different angle of viewing, camera settings, lighting options, material properties of faces, and foreground and background images.

Starting a Photo Album File and Linking to a Design File

Prior to making a photo album file, you need to open a design file or a number of design files. Then you link the photo album file to the design file. Now perform the following steps:

> *1.* Open the file Joypad.des that you constructed, or open the file from the Chapter 5 folder of the CD accompanying this book.

2. Select File > New > Photo Album. A new photo album file is started. (See Figure 7–2.) In the browser pane, there are two options, Materials and Images, enabling you to browse the materials available from the system and the images constructed in the photo album file. Initially, the Images browser is empty because there are no images constructed.

Figure 7–2
New photo album
file started

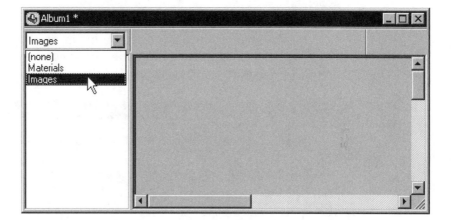

3. Select Image > New Image.

4. In the Choose Modeling View dialog box, select a design file. If there is only one design file opened, it will be selected automatically. (See Figure 7–3.)

5. Select the OK button. An image linked to the selected design file is constructed.

6. Save your file (file name: JoyPad.alb).

Figure 7–3
Choose Modeling
View dialog box

Image Properties

Parameters contributing to a photo-realistic rendered image can be categorized in two groups, image properties and material properties.

Image properties include image size and quality, foreground and background effects, and studio lighting and camera. Now perform the following steps to set image properties.

1. Select the image icon from the image browser pane, right-click, and select Properties, or select Image > Image Properties. (See Figure 7–4.)

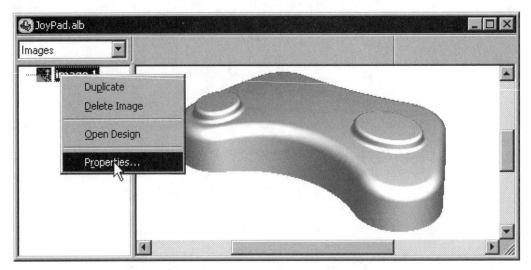

Figure 7–4 Image linked to a design file constructed

2. In the Image tab of the Properties dialog box shown in Figure 7–5, you can rename the image, select a configuration set in the linked design file, set the resolution of the image, and select a rendering quality. Now set resolution to 800 x 600 and set quality to draft. Setting quality to draft saves rendering time. When the settings are finalized and an output is required, you may consider setting quality to presentation.

3. Select the Effects tab.

Figure 7–5
Settings of the
Image tab

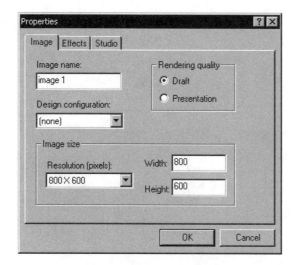

4. The Effects tab has two settings, foreground image to superimpose on the image and background image that the rendered image will be superimposed on. Select None in the Foreground pull-down list box and select Custom in the Background pull-down list box.

5. Select the Browse button.

6. Select the image Flower01.bmp from the Chapter 7 folder of the CD accompanying this book and select the Open button from the Open dialog box.

7. On returning to the Properties dialog box, select the Studio tab.

Figure 7–6
Effects tab of the
Properties dialog
box

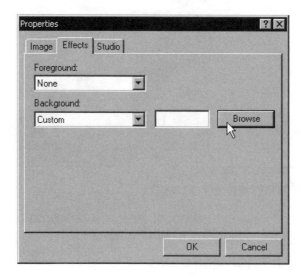

8. The Studio tab of the Properties dialog box shown in Figure 7–7 has two settings, Lighting and Camera Lens.

9. Select Room Lighting and Standard (50mm) lens.

10. Select the OK button.

11. Save your file.

Figure 7–7
Studio tab of the
Properties dialog
box

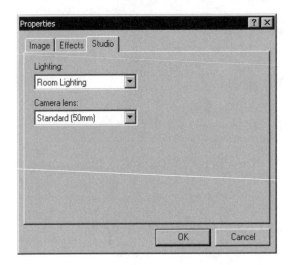

12. Select Image > Update Image. The image is updated. (See Figure 7–8.)

Figure 7–8
Image updated

Material Property

The prime factor affecting the realism of an object is its material property. Now perform the following steps.

1. Open the file JoyPad.alb, if you already closed it.

2. Select Tools > Material Browser, or select Materials from the browser pane.

3. Select the solid part.

4. Select plastic polished from the material browser, right-click, and select Apply Material. (See Figure 7–9.) Material property is applied.

5. Select Image > Update Image. The image is updated. (See Figure 7–10.)

6. Save your file.

7. To output a rendered image, select File > Export > Tiff (or other formats listed).

8. The image is complete. Save and close your file.

Figure 7–9
Material being
applied

Figure 7–10
Material applied

Constructing Animations

Constructing photo-realistic animations also requires the construction of a photo album file. Animation refers to the animated configurations of the linked design files. Therefore, animation can only be constructed if there are configurations set and animation established in the related design file. Prior to generating an animation, you also need to set image and material properties.

Now perform the following steps to construct an animation.

1. Open the design file PlaneS1.des that you constructed or open it from the Chapter 5 folder of the CD accompanying this book.

2. Select File > New > Photo Album.

3. Select Image > New Image.

4. In the Choose Modeling View dialog box, select PlaneS1 and select the OK button.

5. At your discretion, set image properties to the image and set material properties to individual components of the assembly.

6. Select Tools > Animation.

7. Check the Movie output file box and specify an AVI file name in the Album Animation dialog box shown in Figure 7-11.

8. The image is complete. Save and close your file.

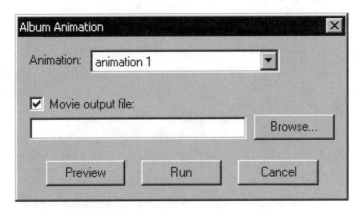

Figure 7-11 Album Animation dialog box

Exercises

Now enhance your knowledge by working on the following exercises.

Ballpoint Pen Project

Construct a set of photo album images for the components and the assembly of the ballpoint pen.

Toy Plane Project

Construct a set of photo album images for the components and assemblies of the toy plane. Generate an animation for the toy plane assembly.

Summary

To enhance visualization and presentation of the 3D objects depicted in a design file, you produce photo-realistic rendered images and animations.

Making photo-realistic rendered images and animations requires a photo album file, which links to one or more design files. While the design files depict the geometry of the objects, the photo album file stores two sets of information about the 3D object. The first set of information regards image properties, including image size and quality, foreground and background effects, and studio lights and cameras. The second set of information concerns the material properties of the faces and parts of the linked design file.

After setting these properties, you can output photo-realistic images from the linked design file. If there is a set of configurations and an animation of configurations defined in the linked design file, you can output an animation.

Review Questions

1. Can a photo album file be linked to multiple design files?
2. Can several images be defined in a photo album file?
3. What kind of image properties can you define in an image?
4. How can you apply material properties to selected faces and parts of the linked design file?

Index

A

About Pro/DESKTOP, 13. *see also* Help systems

Activate Sketch option, 61. *see also* Modification of solid features; Sketching

Add Material option
in the ballpoint pen project, 39, 54
combining solid parts and, 21

Align condition, 145. *see also* Assembly model

All round welding, 240. *see also* Weld symbols

Angled method of constructing a workplane, 28, 29. *see also* Workplanes

Animation
description, 184–186, 253
photo-realistic, 245, 252
in the toy plane project, 198–199, 200
viewing, 9

Annotations to an engineering drawing. *see also* Engineering drawing
centerlines, 226–228
description, 203, 220–221, 244
dimensional information, 221–225
geometric tolerance, 230–234
limits and dimension tolerance, 228–229
notes, 225–226
part reference balloons, 243
parts list, 241–242
surface texture symbols, 234–238
weld symbols, 238–241

Appearance tab, 8. *see also* System settings

Arc button, 33–34

Arc tangent, 33–34

Assembly drawing, 221. *see also* Engineering drawing

Assembly hierarchy. *see also* Assembly model
of the ballpoint pen project, 150–155
description, 136–137, 162
design approaches to, 140–141
of the toy plane project, 155–159

Assembly model. *see also* Solid parts
of the ballpoint pen project, 150–155
check interference tool and, 148
concepts, 135–139
description, 1–2, 14, 135, 161–162
design approaches to, 140–141
design file and, 5
engineering drawing of, 2
generating a 2D engineering model from, 208–220
interfile dependencies and, 9
manipulating components in, 142–144
mating conditions and, 145–148
modifying in context, 159
product structure of, 149
rendered images of, 245–246
of the toy plane project, 155–159, 160–161, 199–200

Assembly modeling task, 138–139. *see also* Assembly model

License Agreement for Thomson Delmar Learning